Nebras Alqurashi is an OT Cybersecurity Subject Matter Expert and Senior Cybersecurity Engineer with eighteen years of experience working with leading OT Cybersecurity vendors worldwide, with expertise in IT/IoT/OT Cybersecurity.

He is an author of OT cybersecurity training courses (OT Cybersecurity Fundamentals and OT Cybersecurity Professional), as well as a trainer and certified instructor for several vendor-specific cybersecurity certifications.

Extensive field experience providing cybersecurity solutions for OT around the world.

It is my mother who taught me the first letters and mentored my soul throughout my life; it's my father, the best friend, who is always there for me; it is my wife who gives me motivation and believes in me; it's my children who give me pride; it is my siblings who supported me; it's my relatives and friends who always help me along the way.

To those who taught me and mentored me through my career and those who did not ask for anything in return for believing in me and giving me hope.

Nebras Alqurashi

HOW TO BE OT CYBERSECURITY PROFESSIONAL

Austin Macauley Publishers
LONDON * CAMBRIDGE * NEW YORK * SHARJAH

Copyright © Nebras Alqurashi 2023

The right of Nebras Alqurashi to be identified as author of this work has been asserted by the author in accordance with Federal Law No. (7) of UAE, Year 2002, Concerning Copyrights and Neighboring Rights.

All rights reserved. No part of this publication may be reproduced, stored in a retrieval system, or transmitted in any form or by any means, electronic, mechanical, photocopying, recording, or otherwise, without the prior permission of the publishers.

Any person who commits any unauthorized act in relation to this publication may be liable to legal prosecution and civil claims for damages.

The age group that matches the content of the books has been classified according to the age classification system issued by the Ministry of Culture and Youth.

ISBN 9789948789208 (Paperback)
ISBN 9789948789215 (E-Book)

Application Number: MC-10-01-0768166
Age Classification: E

First Published 2023
AUSTIN MACAULEY PUBLISHERS FZE
Sharjah Publishing City
P.O Box [519201]
Sharjah, UAE
www.austinmacauley.ae
+971 655 95 202

Over the course of my career, I have had the privilege of working with exceptional leaders and colleagues who believed in me, helped me and inspired me. I am grateful to all of them.

My goal with this book is to complete the journey of the hundreds who shared their knowledge of OT online, by providing a practical guide to OT cybersecurity.

It is my hope to make learning OT cybersecurity less challenging than it was for me a few years ago when I was unable to find adequate resources.

Table of Contents

Introduction	11
How to Read This Book?	13
Who Is This Book For?	15
Part 1: Fundamentals of OT	**17**
Chapter 1: The Mission	*19*
Chapter 2: Automation	*25*
Chapter 3: Distributed Control Systems (DCS)	*39*
Chapter 4: SCADA	*47*
Chapter 5: Operational Technology (OT)	*56*
Part 2: Architecture and Communications	**65**
Chapter 6: OT Network Architecture	*67*
Chapter 7: IoT and IIoT	*87*
Chapter 8: Introduction to OT Network Protocols	*95*
Part 3: Common Industries	**131**
Chapter 9: Electrical Power Substations and IEC61850	*133*
Chapter 10: Oil and Gas (O&G)	*164*
Chapter 11: Building Management Systems (BMS)	*180*
Part 4	**195**
Chapter 12: OT Cybersecurity Standards and Regulations	*197*
Chapter 13: Risk Assessment	*211*

Chapter 14: OT Cybersecurity Controls 217

References **238**

Introduction

When I needed to learn about OT Cybersecurity in 2017, I didn't find suitable sources at the time. There must be some good sources, but I was unlucky in finding them, and when I tried to find training courses to learn, I found them too expensive compared to the course's outline.

I had to read articles and whitepapers from many sources, watched many YouTube videos, attended short courses on PLCs and HMIs, and started to analyze OT protocols. I am in debt to every person who published their knowledge online for free that I benefited from, and they would be in thousands. I can't list them, and also, working with Stormshield Firewalls helped me a lot because they build protocols plugins similar to Wireshark dissectors that made it easier for me to analyze and understand protocols.

The journey was long and tough, for example reading RFCs to understand protocol specifications was not easy. Eventually, I built a sound foundation for an understanding of OT Cybersecurity.

Since I have been working for many years in securing OT businesses, I have seen the reality, and I know what is written on paper and what is actually on the field. I know about many problems and challenges. I have taken the lead on large projects of securing most critical operations, mainly in the Middle East and Africa, minor in Europe.

This book summarizes my journey of learning OT Cybersecurity. I don't speak the academic language, but I tried in this book to simplify the concepts and link subjects together for you to be able to master the subject and become an OT Cybersecurity professional quickly.

At the same time, this book was written for training. You can consider it as a training class, split into four parts, part 1 establishes the foundation of concepts about the OT Cybersecurity; part 2 explains the communication protocols and OT network architecture; while part 3 showcases the most common industries' example and deep dive into each one of them, and finally part 4 guides you on how to act as a consultant and OT Cybersecurity professional to advise of the best solutions of OT Cybersecurity.

How to Read This Book?

Be patient! Each chapter, either long or short, begins by establishing some concepts. These concepts are essential to understanding the rest of the chapter, so that you will experience the introduction of several subjects at the beginning. Then you will find them in a more relative context altogether by the end of the chapter.

This book is not a reference for protocols, specific industries, or standards. It will take you through the concept, drill down into the essential details, and extend the discussion when required on particular topics.

When OT Protocol is discussed, it will show captures taken from Wireshark of how the packets are formed, and it's essential to get used to that. In the end, in OT Cybersecurity, you are more concerned about understanding the operation and associated risks based on the traffic going through the network, and it's essential to have a sound understanding of the most common OT protocols.

Some protocols cover too many details, and some show less, depending on the need and my observation from actual use cases. However, it only trains you to read and understand a protocol, so following the book will prepare you to have good experience in analyzing other OT protocols that are not mentioned in the book if you need to learn more about them.

We are not looking at packets in real life, while systems and security controls do. Still, we might need to investigate some findings in some cases, leading to the need to inspect the packets manually. Understanding how the protocols communicate will provide a better visibility of how the system works.

This book will not cover the subject of Incident Response (IR) or threat hunting; for that reason, the security controls are covered only for those we expect to minimum have them in operation, and that's the reason we don't cover much of, for example, SIEM, SOC, or EDR instead, we highlight the need for archiving of system logs and audits only because I do believe the OT IR training is a different advanced subject and might be a candidate for future book/training. Still, this book is the prerequisite to learning and pursuing a career in OT IR.

Who Is This Book For?

For IT Cybersecurity professionals who would like to learn about OT Cybersecurity, this book is the best fit for engineers and managers alike. Although some managers might not be interested in the details of protocols, they can skip the parts that dive into details of the protocols that are always at the end of the chapters.

Those who are coming from OT/ICS/SCADA background and would like to learn about OT Cybersecurity can skip part one and begin from part 2; however, always a knowledge refresh is good to have. But it's mandatory to have a basic understanding of TCP/IP networks. If that's not the case, I recommend starting from there before reading this book.

The same goes for students or those who are starting their careers, the book would be helpful, but you need to have a basic understanding of TCP/IP communications first.

Almost every chapter is independent, except for Part 1, which is a foundation understanding for all chapters; still, since it's in a training format, some details in one chapter won't be repeated in the next ones. That being said, if you are reading the book only for a specific subject(s), you might be missing some part that was covered in the previous chapter(s), and it makes no sense to repeat it in the same book, so you may need to at least read all chapters within the same part your subject is listed within.

Finally, I hope my life experience and knowledge that I have put into this book can be informative and helpful for you, and I sure welcome all comments and feedback you may have. Feel free to connect with me at any of the following:

nebras@otsec.io
Website: www.otsec.io
Twitter: @NebrasAlQurashi
LinkedIn: Nebras Alqurashi

Always check www.otsec.io for updates, sample files for analysis, new articles, and short videos and training programs.

Part 1: Fundamentals of OT

Chapter 1: The Mission

While going over social media platforms, watching people of different ages from all parts of the world posting funny and sometimes ridiculing stuff, rich and poor alike find their way to the internet, sharing and debating on any subject. It's mind-blowing to uncover how much technology plays a significant role in our lives. You get the sense of the necessity to have access to the internet, comparable to having water, food, and shelter!

The existence of the internet resulted from technological evolution over the years. Since the invention of computers, technology has found its way to get things connected. It started with very simple goals and ended up in billions of connected devices all over the globe!

Operational technology, similarly, is an outcome of the evolution of the machine! It's where things act to serve the purpose they were designed for. It is run by some logic that informs machines how and when to work. They could be a simple printer or lighting system; others are more complex, like chemical manufacturing or power generation and distribution.

As an OT Cybersecurity professional, you need to understand the system you want to secure. You need to maintain a level of knowledge around it, which helps your mission in securing it while staying aligned with the business and operational objectives.

For example, for an IT Cybersecurity professional who desires to implement a network security control, the minimum required knowledge is to understand the basics of networking and how the systems are interconnected as well as what type of switching and routing is in place. Furthermore to understand the risks and what must be eliminated or reduced to an acceptable level, after that it's possible for the IT Cybersecurity professional to design the network security solution and implement the required controls. Similarly, for OT Cybersecurity professionals, it's an obligation to have a good level of understanding about the operation to be secured.

Another example is the Web. Those who are experts in web security and web penetration testing know all about XSS and SQL injection, but not necessarily how they can program or design a website. They know how the web services are designed and how they are programmed up to a certain extent, only enough to be able to secure them or break them. If you are not coming from an OT background, then you need to obtain a solid understanding of the fundamentals of OT, what the components are, and how they communicate.

For OT Cybersecurity professionals, it's also essential to understand your mission. You will learn skills over the course of this book, but at the same time, you need to know what the objectives are; besides how you're doing it, you may need to think about why you're learning and what could be your mission in the future in your current role or the role you are planning to acquire.

What is essential in this chapter is to grasp and recall the historical events around the Industrial Revolution and understand how it evolved along with its impact, which will make your mission clearer; then, you can live up to the good cause of securing it.

It all began in the latter half of the eighteenth century for the agricultural societies in America and Europe moving toward urban industrial productions.

The invention of steam power was game-changing, first in Britain and later spread amongst the rest of the world; this invention has changed the business model from manual and hand-crafted products into mass quantity production.

The 1830s and 1840s are the periods of what is called later the First Industrial Revolution (Industry 1.0).

Followed in the late nineteenth century by the Second Industrial Revolution (Industry 2.0) that had rapid advances in the steel, electric, and automobile industries.

When we watch over the revolution of the industrial world, we can observe the changes to societies and culture; the textile business, for example, that used to be made in small workshops by individuals, has shifted to mass production manufactures, that had an impact on individuals' lives, laws, economy, politics, and many others.

When you are on the journey of learning OT Cybersecurity, you must understand why we are trying to create resilient and cyber-safe operations; you will realize the impact of the industries on our lives.

Let's take the steam power; it was initially invented to pump water out of mine shafts. It was then improved to rotary motion, that had contributed to many industries like flour, paper, cotton mills, ironworks, distilleries, waterworks, and canals.

It was not only more efficient and contributed to many more industries but also has increased the demand for Fuel. At the time, it was coal, which led to creating opportunities for miners to go deeper and extract more of the cheap coal that created the demand for better transportation and distribution, with all the impact on the culture and work opportunities will arise.

The Telegraph was invented to answer the demands of quick long-distance communication, followed by the steam power revolution. Eventually, the need for a banking system to cover commercial transactions.

Many farmers felt the need to move to main cities to secure jobs in the new work models, making cities overcrowded. This rapid urbanization brought significant challenges, as crowded cities suffered from pollution, inadequate sanitation, and a lack of clean drinking water.

We can quickly realize the impact of the Third Industrial Revolution (Industry 3.0) when information technology and electronics were used. Processes are automated and operating with almost no human interference; it has shifted the human work to more supervisory than manually. It has evolved over the years, improving efficiency and safety.

As Industry 3.0 has witnessed the age of the Internet, IT components have increasingly played a substantial role in the operations. Internet initially was there for simple ideas, how to print from different workstations to one printer, and it has evolved as we are experiencing today. The same was applied to Industry 3.0, components of an operation became things talking to each other, the type of "talking" has also evolved from one-to-one serial communications to standalone communication protocols over Ethernet networks.

Speaking of history and the detailed evolution of the industry is a large subject. In today's reality, many events have occurred over hundreds of years, where we are still witnessing Industry 3.0 technology and an urge to shift toward Industry 4.0. In the following chapters, we will discuss the components of operational technology in detail. You will understand how technology looks today and how it will be in the near future.

Fourth Industrial Revolution (Industry 4.0) is the natural evolution in the age of Big Data; considering the tremendous amount of data in operations that are processed in a second, over time, this information can be utilized to improve the operation from different aspects such as efficiency, safety, security, maintenance, economy, planning, and others.

If you have the information for the last six months about an operation, you can build different views using Artificial Intelligence applications to investigate other cases; what if you correlate the information about a process for a longer duration along with the information about system maintenance for the same time, wouldn't that give you an idea of what could be the reason for frequent failures of system parts? It could, and it's one of the objectives for maintaining the data of an operation.

What if you could have insights from similar operations worldwide loaded onto cloud-based intelligence? It will create a rich data lake of information about the specific process, continuously being analyzed for many use cases. In return, you can obtain better guidelines and recommendations on optimizing and achieving better and more efficient operations.

The simulation used to be essential for training, but also with Industry 4.0, we have the concept around Digital Twin, where you can have the same physical operation simulated in virtual systems that can tune even the environmental changes to simulate temperature, humidity, vibration, and many other factors to study the impact and, of course, to test different logics and optimization without the need to interrupt the actual physical system, this will lead to great results with minimal impact.

Systems have their languages and an interface to speak to; designing an app that can talk to an API of another system will take the human off the formula and let the systems talk to each other and do the job for you.

We have smart cars, smart buildings, smart homes, smartphones, smart factories, smart things. And we are taking that for granted and enjoying new tech! Unlike the traditional concerns that machines will take over human work, it has improved our quality of life and at the same time created more work opportunities, but less handy while techier!

Human cultures have changed, and you can see teenage collectively behaving likewise and interested in similar subjects regardless of which part of the world they are. That's off subject, I know, but what we need to observe is the fact that systems are becoming more interconnected, even if not directly or at least dependent on each other.

If you recall any movie about the end of the world and apocalypse, they usually show how our life will look like after the knowledge we obtain in industry 4.0 and then how to live off the grid and try to live by simple tools; these movies begin with a significant or small event that led to series of events ended up in the way of living in the wild world!

Real wars in history or those we're witnessing in our lives today would also show similar scenes and maybe more brutal than what we watch in movies.

The truth about today's life is that it depends on the industrial part of the world, food supply, power generation, transportation, and water supply; they are all essential for living while any interruption may occur. It will have a severe impact on our social life order.

Personally, it amazes me how much our lives today depends on systems and it scares me how these systems are vulnerable to cyberattacks! This is the mission you need to take seriously if you're pursuing a career in OT Cybersecurity; it's like many other careers that impact human lives but on a larger scale.

It's also important to realize that OT Cybersecurity is a nations' defense system; with more political conflicts globally, we observe numerous cyber wars. The motive is not for fun or more money in their pocket. These attacks have targeted the safety system with the clear objective of taking human lives. You may be wondering, why not keep using legacy systems to mitigate the risk?

Digital Transformation

We are witnessing an ERA of digital transformation in both worlds of IT and OT; it's pretty required and justifiable for IT, it's the norm of IT to adapt to new technologies and trends constantly, but within OT, it's usually slower. The OT community takes a long time studying and deciding on such a move.

Considering the risks, we have discussed earlier, that might make you wonder why we expose OT to more risks by making it rush into an adaptation of advanced technologies? Why not keep using legacy systems and remain safe and secure?

To answer this question, I will first list some critical challenges with legacy systems as follows:

- Many legacy systems are reaching end of life. The maintenance cost is extremely high; some parts may not be available anymore, making it impossible to keep the operation running for the near future.

- The consumer market is not the same as it used to be ten years ago; in some industries, it's impossible to continue without change, which might not be possible to reach the market's expectations.
- Regulations and compliance are very challenging for legacy systems.
- Environmental impact and power consumption are worse at legacy systems.
- It's an open market, and operations are businesses, and they need to be competitive in quality and cost; they cannot stay in the game if they run on legacy systems while their competitors use new efficient digital systems.

New digital systems would bring many benefits to operations, answering the above challenges of legacy systems. It also boosts the overall performance and utilization; for example, it may increase 10–20% of sales, equipment utilization, and cost reduction.

Digital transformation is a needed journey, and it has made a significant contribution to our lives! No operation can swim against the stream and stay in business, it must be acknowledged, and your mission on this journey is to make sure the transformation is happening safely and securely.

I hope this overview of the Industrial Revolution and Digital Transformation helps you get insights into the need for your mission. Considering the world today is in shortage of OT Cybersecurity skills, you can fill the gap while securing a promising career and having a good cause to live up to.

Chapter 2: Automation

Automation is a technique to make a process or system works automatically. And as per ISA, "International Society of Automation," it's the *creation and application of technology to monitor and control production and delivery of products and services*.

As the first chapter indicated, over time, we "Humans" develop our tools to make our life easier; our objective is to automate everything, not only in industries but in all aspects of our lives.

I trust many automation tools in my daily job that can let me work with fewer efforts and be more productive; I have friends obsessed with python automation. They have created simple programs that can do some repetitive tasks, which I admire.

The concept of automation in the industrial world is not a simple process. Still, it's a collective set of processes and components that simultaneously work to achieve a specific goal, like manufacturing a product.

Let's discuss an example of juice production; what are the different processes involved in such an example?

Juice itself requires mixing and processing several ingredients to produce juice; you need to supply and store these ingredients, then bring them into the mixer with exact amounts and accurate mixing technique and timing, while maintaining the hygiene level without impacting the taste (quality) and health. You store the produced juice in safe storage, later filling an exact amount into containers.

The container itself needs another process to cut cartoons and press them into the shape of a product while maintaining quality and health guidelines. It also requires labeling or printing before shaping or after (such as production/expiry details and serial numbers). Supply and storage of empty products is another process that must be maintained.

You can have part of these automated processes called "Semi-Automated," where a portion of the work cycle would be done manually, and the other part would be automated.

Another type is that the whole process can be automated and called "Fully Automated," which means the processes will operate for an extended period without human attention.

Automation by itself is not an objective; it's a way to achieve the aim of production; we need to realize that many factors play significant roles in deciding what type of automation is required. If the cost of acquiring, operating, and maintaining an automation system for a process or part of a process cost more than manual work, then it's not desired. At the end of the day, in our manufacture example, they are running a business. If they need to buy a multi-million $$ packaging and labeling system that is not feasible for their business case, they might rely on manual work. However, that's to provide you insights why we may still have semi-automated systems.

We would have fully automated systems in the ideal case, but there is more to realize in this part. Let's consider our juice production example fully automated, which means we need to look at it as several processes. Each process has inputs and outputs, a duration of time, and an order in the which process comes first and comes second until last.

That means during the design phase, all these processes must be analyzed separately to identify the objects, like:

- Input (raw material)
- Speed (how many products it can process at a time)
- Full cycle time (how long it takes to process a single product or batch)
- Output (processed input)

Then it would help if you understood what might go wrong:

- Quality
- Quantity
- Risk on health
- Risk on labor
- Risk on environment

Accordingly, you will need to have processes to have a safe shutdown or avoid the error from being repeated on the following processed product.

Having each process analyzed separately, you can then have the proper design and fine-tuning of all processes to work together, considering the different natures, risks, and speed. Eventually, you have the design ready for how these processes will work together jointly to produce a product.

Another review of risks and safe shutdown of an overall process, which means considering a single process would be different from multi-processes working together, and even in the case of an emergency shutdown, you need to follow a procedure of what to be shut down first and what to be delayed.

This example is to give you an idea of how automation is being designed; in reality, it has more details, and automation is a science that is being taught in universities for years; in the end, this example is not enough to start your juice factory, that's for sure!

Now that you have an idea of how automation looks for simple manufacturing, imagine how it will be the case for petrochemical and pharmaceutical industries, oil & gas, or even power generation and distribution. The nature of sectors is different and has specialties; in a power grid, if you look at the process of a power substation, it's concerned about transmission, elevation, and reduction of the voltage of thousands of Kv. The processing speed is different from a process that monitors the flow of oil pipeline between two regions.

Automation is a broad subject, and the takeaway as an OT Cybersecurity professional is that you need to have adequate knowledge and not as an expert in automation, at least to be familiar with general industry processes and components. However, we will discuss the details of the most important ones at an adequate level.

We can classify automation by the possibility to make a change, where we have three basic types:

- Fixed Automation: it's impossible to modify the process in this type.
- Programmable Automation may apply changes to the program it's designed for, but you cannot change the functionality.
- Flexible Automation reduces setup time and can switch to different functionality.

Setup time is the interval needed to adjust settings for the process to be ready to run, whereas run time is the interval required to process a single unit.

By having this common knowledge about automation, you need to know how it's evolving, like everything else in the ERA of digital transformation.

Automation must be converted to machine logic to make real things work by the logic we have created. In OT, it's where you have the parts that do the work, or we call them "Actuators" and the parts that monitor and provide us with information "Sensors" and the logical part that has all the logic stored and processing the information from sensors and providing work orders to the actuators, "Control Logic."

The control logic is read in the form of a program, this program is similar to IT programming languages, but we have particular programming languages in automation.

The program will run in a cycle, from start to end, then again. If we don't have this cycle loop, we wouldn't have complete automation, which means the process runs once per processing unit. Once done, it runs again for processing another unit and so forth.

When you have advanced processes, the program relies on several readings from sensors at different times to send commands to the actuators to work. The complete program cycle might have several readings from "Input" sensors and commands several actions "Output" to actuators.

An example of a sensor is to read the temperature, pressure, distance, and voltage. They may use microwaves, laser, or ultrasonic technologies, where actuators might be a robotic arm, drill, and produce heat, energizing motor, or pump.

Water Tank Example

In this Water Tank Automation system, we have the following components:
Active Components:

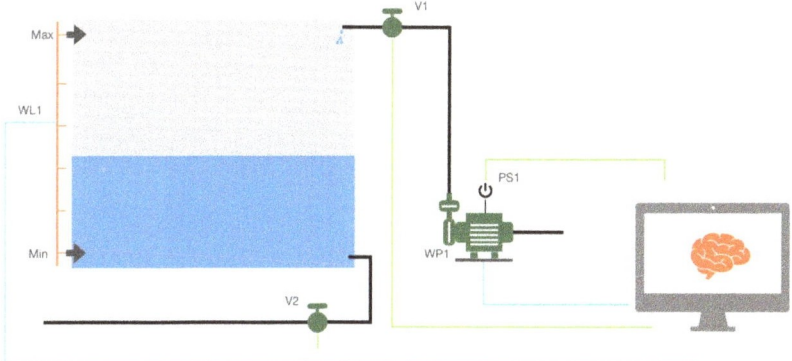

- Valves (V1 & V2), simple actuators that can be in two states (open/close)
- Water Level (WL1), A sensor that provides a water level reading in a tank.
- Water Pump (WP1), An actuator that pumps water from the source into the tank.
- Power Switch of Water Pump (PS1), Power Relay an actuator that switches power to (on/off).

Passive Components:

- Tank.
- Pipes.
- Power lines.
- Water.

Here we have distinguished the components into Active or Passive; the Active components can be part of the Logical Programming of the Automation System, whereas passive components are not measured and have no representative value within the Logic.

We have assigned active components variable names (V1, V2, WL1, PS1, WP1) because we need to either read their values or set new values in the program.

These active components are also classified as either sensor or actuator. Sensors are always there to provide us with values, so they can also be referred to by "Input" or "Read Only," whereas actuators can be referred to as "output" or "read/write."

Sensors can provide analog values; for example, the water level sensor won't provide the exact level of the water in the tank; in fact, it gives an analog value based on a scale from min to max, and later in the automation system, it will be required to perform scaling or conversion to human-friendly language.

In our example, the water level sensor will be providing an analog value between -789 to +789, where 0% is represented by -789, and 100% is represented by +789. And when you have a zero analog reading, it means 50% tank capacity.

This conversion is part of the automation programming and handling of different sensors and actuators. Due to these devices' different natures and options, they will be using different scaling, which needs to be considered in the programming part.

The control logic is represented by the brain, which, as you can see, all active components are connected to it. We have two types of connections wherein our example, we have sensors connected through blue lines and actuators in green; in this case, we can say the control logic has two types of connections Input and Output.

What might be confusing is why I put WP1 considered a sensor on purpose, although it's an actuator? The reason for that in the automation, we're only reading the state of the pump if it's On or Off, where the energizing is occurring at the power switch PS1.

Now let's go through the logic of this example, and it makes it simple to consider that we have an endless water supply, and water can flow out to the next phase endlessly as well, in order not to worry about checking before or after the process.

The logic:

Check the level of the water tank by reading sensor WL1, and we have the following possibilities:

- If the level is less than 10%, we need to stop flowing water out of the tank and pump water into the tank.
- If the level is less than or equal to 90%, we need to pump water into the tank and let the water flow out of the tank.
- If the level is more than 90%, we need to stop pumping water and flow water out of the tank.

Take a moment to read the following logic while tracing it on the diagram:

- Stop water flow from the tank is performed by closing valve V2.
- Flow water out of the tank is performed by opening valve V2.
- Pump water to the tank by opening valve V1 first and then energizing power switch PS1.
- Stop pumping water to the tank by first shutting down power switch PS1 and then closing valve V1.

Order in some actions is usually mandatory for the system's health and safety in general.

This logic will be written in a program format to the control logic and run once every 1 second.

Someone supervising the system will have a screen that shows the values of different variables, or let's refer to them by "Tags" in real-time.

At this level, we understand how the automation of such a system is designed; now, we need to make it more realistic and put that one process along with other processes.

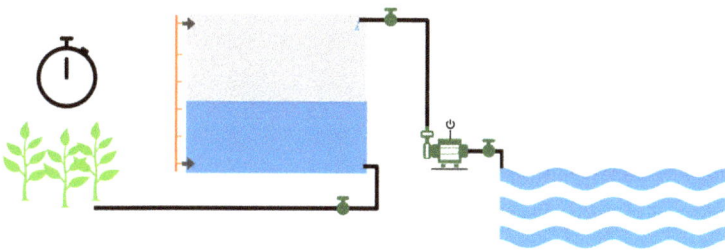

Our designed process will fit into the model of other processes; we have at first one process to verify the availability and maybe quality of the water source. Then we have another process to control the flow because the water will be used for plant watering and run per schedule.

We will not go into further details, but at this stage, I believe the idea of automation should be more convenient to you by now.

What happens if any of the system components fail?

If the water level sensor is broken, it will cause a severe problem; for that reason, it's common practice to have a safety system in place to take over in case of any component failure.

Safety Instrumented System "SIS"

It's used in critical operations to reduce or avoid the possibility of accidents and injuries; it's another set of sensors and actuators with control logic to check for specific conditions that, if met, will make a safe shutdown of the operation in the predefined procedure.

In our water tank example, let's consider it critical. If it fails, it may cause damage to the environment and may cause loss of lives, so we need to make sure it's monitored and avoid the bad possibilities, and nothing is left for chances.

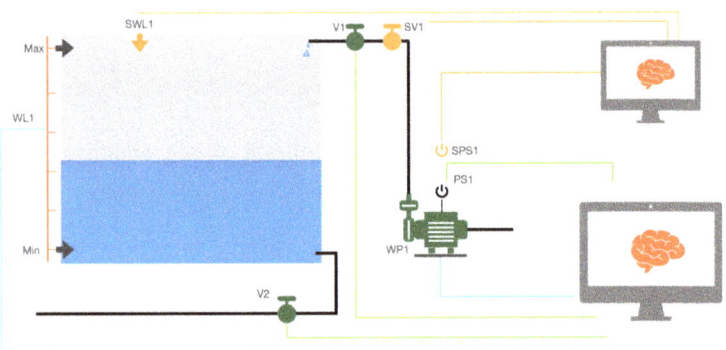

In this case, we need redundant components:

- Logical Program
- Safety Water level (SWL1)
- Safety Valve (SV1)
- Safety Power Switch (SPS1)

This isolated system monitors the same operation; if the valve is broken, it can take over, close the flow, and shut down the power switch. If the reading from the sensor is not correct, it will issue a shutdown procedure. Usually, it will be associated with alarms.

Data Historian

It's the process of continuously reading operation variables (tags) and storing them into the database in real-time; the gathered information can be used for obtaining information about the production, health of the system, detection of failures, and optimization.

It's similar to real-time logs, usually not written to a structured database but more of a flat database.

We have different tags in our water tank example; we have analog values readings, we also have binary "state" on/off and open/close. In general, when tags are read from the operation, they might have different labels like timestamp, quality indicator, limits, and calculation of separate tags into one new tag.

We can have this table of tags and the reading is occurring every 1 second:

Tag name	Data type	I/O	timestamp	Reading	Scaled	Average
ps1	Binary	O	21032022 15:23:18	1		
wl1	Analog	I	21032022 15:23:18	1037	70%	75%
wp1	Binary	I	21032022 15:23:18	1		
v1	Binary	O	21032022 15:23:18	1		
v2	Binary	O	21032022 15:23:18	1		

Tag name	Data type	I/O	Timestep	Reading	Scaled	Average
ps1	Binary	O	21032022 15:23:19	1		
wl1	Analog	I	21032022 15:23:19	1038	70%	75%
wp1	Binary	I	21032022 15:23:19	1		
v1	Binary	O	21032022 15:23:19	1		
v2	Binary	O	210s32022 15:23:19	1		

The table should be easy to read and understand what is going on; we have two readings in the duration of two seconds; the water level is at 70%, which means we can pump water to the tank, and at the same time we can flow water out of the tank, we can see that the water pump and power switch are both ON. The two valves are open, so we have water in and out of the tank simultaneously.

Now look at the following example, and try to spot a problem:

Tag name	Data type	I/O	Timestep	Reading	Scaled	Average
ps1	Binary	O	21032022 23:10:38	1		
wl1	Analog	I	21032022 23:10:38	118	8%	8%
wp1	Binary	I	21032022 23:10:38	0		
v1	Binary	O	21032022 23:10:38	1		
v2	Binary	O	21032022 23:10:38	0		
ps1	Binary	O	21032022 23:10:39	1		
wl1	Analog	I	21032022 23:10:39	118	8%	8%

wp1	Binary	I	21032022 23:10:39	0			
v1	Binary	O	21032022 23:10:39	1			
v2	Binary	O	21032022 23:10:39	0			

The quality indicator is the first value to check; the average water level is 8% in the two readings over two seconds. The standard is calculated based on the duration of time; let's assume it's an average of one hour, which means we have had a problem for 1 hour, or even if we need to look at older data to check for how long the problem has been there, our logic is to stop flowing the water out of the tank when it's 10% and to pump water to the tank.

We can see that valve two, which flows the water out of the tank, is closed, in both readings, and this part of the logic is working fine. The power switch must be on to pump water to the tank, and it's the case, so what is the problem?

We can see that the water pump itself is OFF, which is an indicator of the problem, and it's the reason we have such a low level of water.

Baseline

What to do with all the data we collect and store in our data historian? We can build reports and put conditions to alert us if anything goes wrong. We can create reports and dashboards to overview the operation over time. We can obtain details about the consumption and many other use cases.

We can, over time, understand that if you have a valve that has been open/closed more than two million times, it's broken; this will help plan preventive maintenance.

But from the safety and security of operation point of view, we can understand what readings to expect as an indicator of a regular operation. What are the values that show we have everything functioning appropriately?

Based on the data historian, we can build the baseline of a reliable operation, which is simply having sample values that we expect within the stable operation. Any deviation from these average values indicates we have a problem.

We will see later in this book that it's an essential aspect of **Intrusion Detection Systems** "IDS" that can alert in the case of some suspicious behavior of an operation.

Is it the same for different industries?

We have discussed one example of automation; although the operation itself looks modest, you can now understand that it involves many details when it's automated; how will it look for a power substation or oil refinery?

When we discuss industrial protocols later in this book, we will analyze the main used protocols per industry, and we will go into a deep technical dive for each one of them; however, for now, it's essential to realize when you look at oil refinery operation, you can expect how many values you will need to monitor to understand the regular operation and can spot the deviation from that.

The communication of these tags will be in the mean of industrial protocols over Ethernet networks; in later chapters, we will drill down to the components that communicate the operations controls and monitor the tags.

But to the question of how the automation will be per industry? The concept is always the same; we can have Building Management Systems "BMS" or Building Automation Systems "BAS", which operates a building. It has to monitor and control every physical system in the building, such as HVAC, access doors, parking lots, CCTV, elevators, lighting, power…etc.

Each subsystem within the BMS has its processes and automation and types of tags being read or controlled and modified.

As an OT Cybersecurity professional, what concerns you is monitoring what is communicated over the Ethernet network; the operation is designed to achieve the monitoring capabilities. Still, you will be focused on cyberattacks aspects. Hence, you need to have a common understanding of the operation's objective and how it works, then define what type of regular communication to expect, and then identify anything suspicious to investigate along with the engineers of the operation to either justify, ignore, or respond to an incident.

In manufacturing, it's different in nature from systems like BMS; we have two types of manufacturing:

- Process Manufacturing

It's when change is applied to raw material. Let's say you are making a birthday cake, 'I wish to learn how to make a proper one,' obviously, you will need to cook a mix of ingredients in the oven. However, it's impossible to reverse the process to extract the original elements from the cake after it's cooked!

Process manufacturing is always a one-way process with some inputs processed by formulas or recipes. Then you get processed output that can't be returned to its original raw materials. It uses measurements like weight and volume.

- Discrete Manufacturing

The process of assembling several parts to form a product is most probably since you are reading this book, you have probably tried to reassemble toys when you were a kid.

Automobile manufacturing is one example; it's formed from assembling several parts. Even the elements themselves could be made originally from other sub-parts. Some parts in discrete manufacturing could also be of a process manufacturing type.

It's often measured in pieces, and you can see it assembled in different stages of a production line.

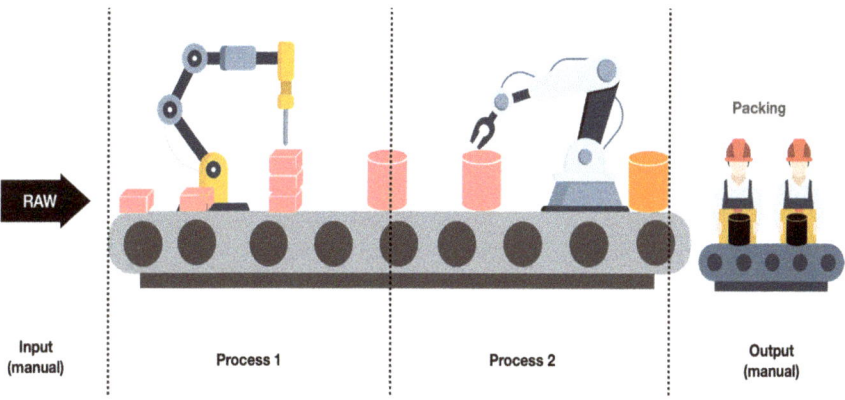

This illustration of a production line is for semi-automated discrete manufacturing. Loading to and from the production line is manual, and the production line has two processes; process 1 mounts several parts to form a product, and process 2 does the packaging and moves the ready product to the buffer area.

In both manufacturing types, they can produce products as one product at a time or in form batches which is more efficient.

In summary, the physical operations can be automated. The logic must be designed as per the operational requirements and turned into a logical program that controls actuators and monitors sensors while maintaining the readings of tags into the historian database.

Safety Instrumented Systems (SIS) will be a separate automation within the same operation to take over control in case of failure.

We have a process and discrete manufacturing; they can be fully or semi-automated and can produce a single or batch of products.

Chapter 3: Distributed Control Systems (DCS)

It is a computer-based control and operation system wherein the automation chapter; we referred to the program logic; in DCS, we speak about the Controller or the computer that stores and executes the program logic and manages the I/O with sensors and actuators.

On the other hand, the DCS has different components, like:

- HMI: Human Machine Interface, it's a monitor that has a graphical representation of the process and offers both monitoring and control of several parts of the process.
- Historian, already discussed in the automation chapter.
- Engineering software, usually found on an engineering workstation, has the software suite to fine-tune the logic of the Controller.
- Common database, related to the business use case of production count.
- Alarm management is a system that notifies the operation supervisor if something goes wrong.

The DCS is an autonomous system; it makes the process or plant run independently from other processes or plants; it has its local controllers to manage process-specific work. The DCS combines several processes into one automated system, including all the above components.

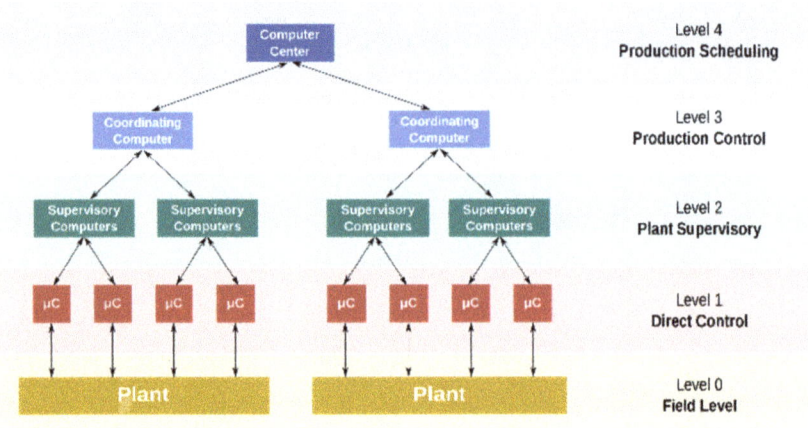

Functional levels of manufacturing control operation

From the above functional levels' illustration, you may observe that sensors and actuators of a plant are controlled directly from "direct control," the direct controls are monitored and controlled from supervisory computers and up through coordinating computer to the central computer that has an overview of all the plants in the operation.

The structural organization of DCS systems makes it the best choice for continuous and complex processes such as petrochemical manufacturing.

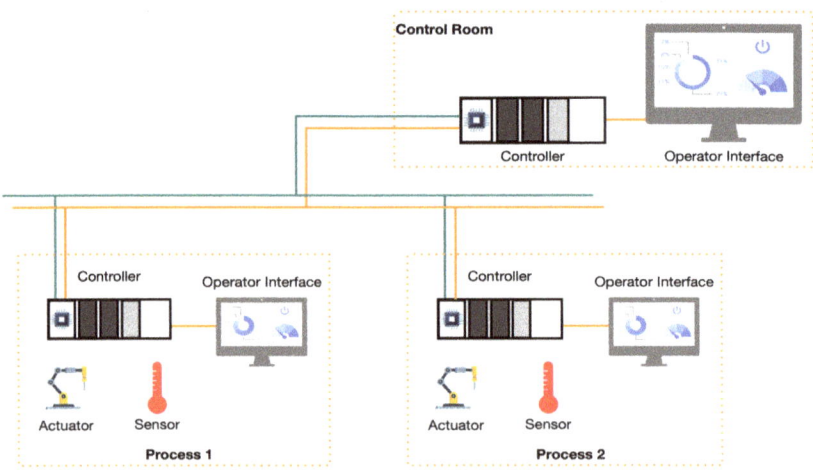

The Controllers have input/output modules; these modules could be for analog or digital values (0/1).

All the work is centrally managed, and at the same time, has local controllers to perform individual processes. If one direct control fails, it has no impact on the other processes and plants in operation.

The control room is usually local and not remote for the DCS system; unlike SCADA, which we will discuss in the following chapter; all the components are connected locally at a high-speed Ethernet network and have a high level of built-in safety.

The DCS systems are provided entirely from the same OEM vendor, so integration between different components is not a challenge to be addressed because it's already integrated as one DCS system.

The DCS systems come with rich functionalities; the engineers will be enabling or disabling features rather than programming the logic from scratch, as it's the case in PLC, which we will discuss in the next chapter.

When you buy a printer, it's a complete system. All that you need to do is to load paper onto it and then send a print command to it, and it will do the job; you can access the built-in GUI (Local HMI) of the printer to make some configuration adjustments, such as connecting it to WIFI or specify date and time. Still, you don't change the printer's program itself, and you don't integrate cartridges from other OEMs into your bought printer; you must use the exact model and vendor to work. The printer is a mini example of a DCS system.

Another example is an electric car, it has many functionalities and an HMI interface, you can adjust settings of the lighting, AC, parking sensors, audio, and other excellent features, but you don't write the code for the car, nor you can integrate parts or systems from different manufacturers.

I put these two examples of products that act as a DCS system because it's important to distinguish the DCS from the SCADA system.

DCS systems are reliable, and safety is a critical system that comes built-in, and it's normal to see many redundant components within DCS for high availability.

The Ethernet communication within DCS is usually high-speed and today mostly over fiber optic, or at least the backbone between different levels is over fiber optic, it's also redundant so you can find two links to each component, the use of IT link aggregation and Ethernet channel is widely used. And it also uses some redundancy routing protocols that we will discuss in detail in a later chapter of this book.

For an OT Cybersecurity professional, you need to have this common understanding of the DCS system and expect to face very complex architectures in real life. Still, you don't need more than understanding the flow of communications and the general role of each component, so always try to simplify the reading of any DCS architecture and create a summary of components and communication flows; we will discuss that in detail in a later chapter. We will also be discussing the logical levels of where each component belongs.

The DCS system is commissioned through different phases and a well-documented project plan; engineers always know which stage they are at, and for each step, they have a checklist to make sure it's being commissioned as per project documentation; for example, they may be checking if the redundant power supply is available and running, the wiring within the cabinet is correct; the earthing is wired; the BOM is matching, the labeling of the modules is correct and according to the proper order.

It also includes phases before commissioning and after the design. Like a FAT (Factory Acceptance Test), which is a test of the system to make sure everything is working and functioning before doing the actual commissioning at the plant.

This usually should not concern you, but you need to be familiar with the process of DCS commissioning because Cybersecurity is being considered part of the requirements, and you might be involved from the early stage of the design or at a later stage.

Importance of DCS systems

To understand the importance of the DCS system, think of a simple process that needs to be replicated 100 times; it's easy to achieve one process and manage it, but when you have a significant number of plants to work on, it's tough without the DCS ability to manage a considerable number of control loops.

For an extensive process like petrochemical, DCS will be the choice because you need to rely on a single integrated system from one OEM vendor that is designed as a whole to achieve the objective of the operation and business.

Now let's give you a feel of how this is connected.

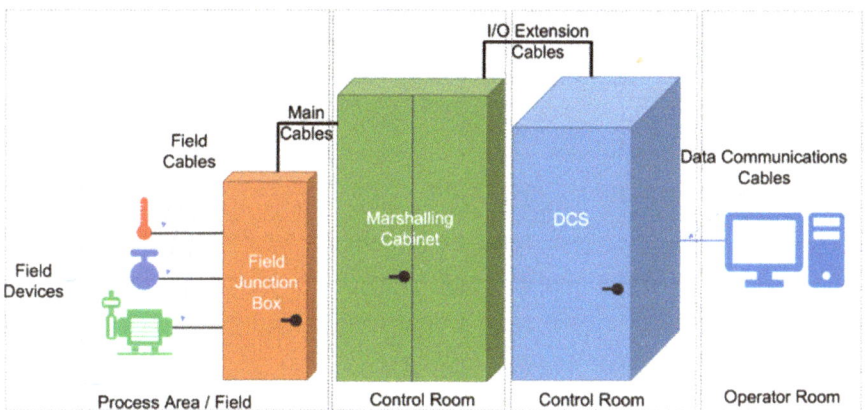

Cabling organization is an excellent quality indicator in any work; you must have seen some before/after cabinet cabling work after the termination to patch panel and organize the cables in an OCD-friendly manner.

Managing the connection cables from field devices will be challenging, usually in pairs, when you have significant numbers of field devices. So, the use of a Junction Box to terminate the field cable pairs and then load them into main cables with several pairs in a single line to the control room Marshalling cabinet.

Marshaling Cabinet is used to terminate main cables pairs and then facilitate the connection of individual pairs to the proper I/O modules in the DCS controllers.

Eventually, through the mean of Ethernet (still you may find some old serial communications) to the Operators' workstations.

PID

PID controller is used everywhere, not specifically for DCS, but also it could be in a mean of part of a program on computerized control. But for DCS, because it has distributed controls, which makes it simple and easier to manage when you have one controller operating one control loop, it's easier to track and troubleshoot, so what is PID?

In the water tank example, the logic was simple and based on a sensor value; we take actions, but what if it's a different liquid or gas? What if it has to be maintained under a specific temperature and pressure? The simple logic of On/Off is not suitable in this case.

The level we desire to maintain of water in the example we call set point, part of the tags that the system reads and stores and also tracks in the historian and other parts of the operation is to know what the set points are.

Now, in reality, you always have a margin around the set point; if I want to keep the room temperature to a certain degree, it won't be accurate because the room temperature will be 1 degree +/- the set point, but how it works?

The thermostat of the AC is a simple example of PID because it relies on feedback from process output; in other words, it has a temperature sensor, and based on the input from the sensor, it either flows cool/hot air or stops, always trying to maintain as close as possible to the set point.

PID is an acronym of **P**roportional **I**ntegral **D**erivative; it's an instrument used to control and regulate temperature, pressure, flow, speed, and other process variables.

It uses a mathematical calculation to identify the difference from the set point and sends commands to tune the process variables to come close to the desired value.

It reduces the error margin to the minimum and improves the quality of the process because it can adjust the actions based on feedback.

Think about a robotic arm or balancing drone flat in the air; it has many variables that can impact the movement, such as the environment or any impact from another part of the system, so you can't rely on simple on/off logic; you need while moving a robotic arm to keep getting feedback of the position to make sure it's going in the right direction or correcting position accordingly.

HMI

Human **M**achine **I**nterface is the graphical interface for operators with the process control systems (DCS or SCADA). It provides a visual representation of the process in the control room; it comes in different modes: single screen, multi-stations, touch screen.

In DCS, the HMI is from the same OEM vendor; it can be an application running on MS Windows or Linux OS and might be embedded or hardened OS.

It can be found connected directly to the controller over serial or connected over the Ethernet network.

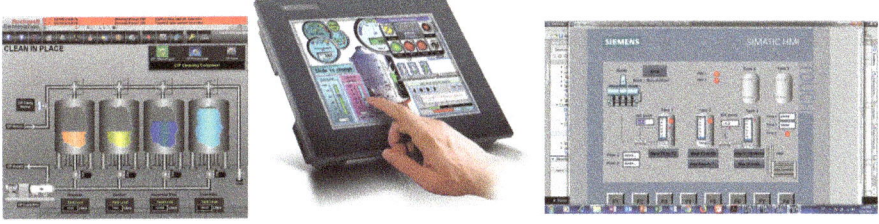

EWS

Engineering **W**ork**S**tation, as per "CISA.org," is usually a high-end, very reliable computing platform designed for configuration, maintenance, and diagnostics of the control system applications and other control system equipment. The system is usually made up of redundant hard disk drives, high-speed network interface, reliable CPUs, performance graphics hardware, and applications that provide configuration and monitoring tools to perform control system application development, compilation, and distribution of system modifications.

From EWS, you can perform firmware upgrades, program logic, I/O configuration, design HMI, and general administration.

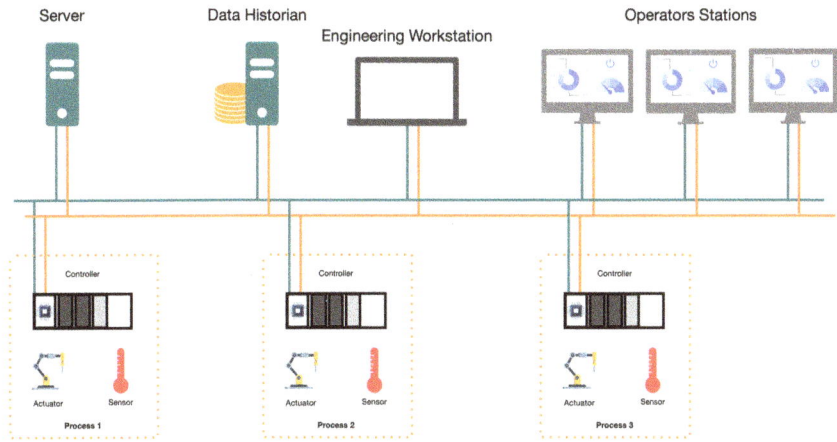

This is a quick overview of the DCS systems; understanding the components of the DCS and a general idea of what purpose they are serving is all that you need to acquire; over the following chapters, we will be referring to these components. Throughout the examples we will discuss, you will feel more comfortable with these components and terms.

Chapter 4: SCADA

SCADA is an acronym of **S**upervisory **C**ontrol **A**nd **D**ata **A**cquisition; from the name itself, you can identify one significant difference with DCS, where DCS is process-driven, the SCADA system is more focused on data acquisition, or event-driven.

SCADA is a system consisting of hardware and software to control processes; these processes can be local or remote. Through monitoring, it continuously gathers process data in real-time.

Unlike DCS, SCADA needs to be programmed from scratch. The system and components will be designed and loaded with the required functionalities and programs, and it's more flexible in integrating different parts from different OEM vendors.

SCADA works on slow internet connections for remote locations; monitoring the processes can identify an error and provide details on how long the error existed. It can send commands to remote operations to correct the errors.

Sometimes SCADA is called Remote-HMI. To indicate the primary use case where SCADA monitors and manages large geographical operations and remote processes. Suppose you want to monitor oil pipelines or power distribution. In that case, they are usually found in a wide area and sometimes across countries, and SCADA is the right solution for such operation.

SCADA is essential for many industries, which maintains efficiency and manages the notifications of any system issues to mitigate downtime.

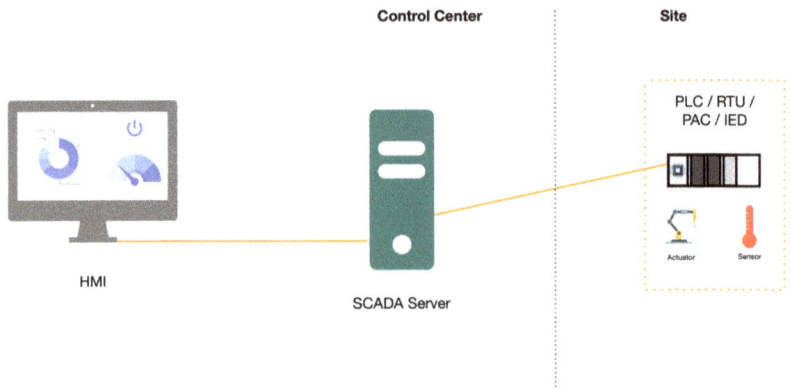

SCADA's main components are HMI, SCADA Server, the controllers (PLC, RTU, PAC, IED), and alarms.

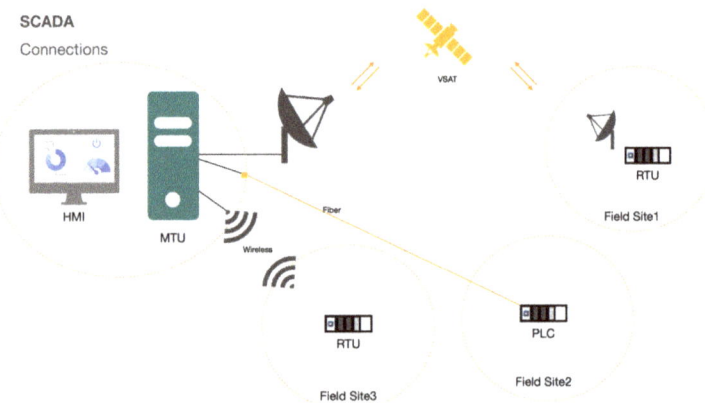

SCADA Server

Main communication point between the control center where the SCADA server is located and remote locations. It's also called an **MTU** (**M**aster **T**erminal **U**nit), which indicates the communication capabilities with remote locations.

SCADA Server will collect information from field controls (RTU/PLC/PAC/IED) and provide control commands to these units for error correction and optimization.

The communications of the SCADA Server with field controls are Master/Slave communications (Client/Server), the SCADA Server is the Master, and field controls are slave devices.

This type of communication can work on even low bandwidth. In case of a connection interruption, the operation will continue to work until the connection is restored, and eventually, the SCADA server will resume monitoring the field.

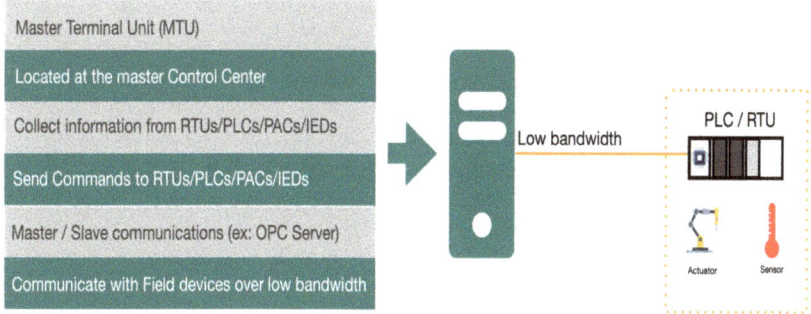

PLC

Programmable **L**ogical **C**ontroller. PLC is an industrial computer that has interfaces with field devices; it may come in a modular structure where several I/O modules can be added to communicate with field devices and execute control loops.

It's programmable, which means the control loop must be programmed in one of the PLC programming languages, and the program is then loaded onto the PLC, which then will be executed. The control loop will run on its own; even if there were an interruption to the communication with the supervisory control center, it would continue to function.

PLCs have network interface to communicate with MTU (SCADA Server). And it comes in different specifications in terms of available resources; some high-end PLCs can manage a significant number of control loops and connect to hundreds of I/O devices.

RTU

Remote **T**erminal **U**nit. Similar to PLC with some differences in capacity. Usually, it's used as a data concentrator for small fields. It has identical functionalities to PLC, but most often, it comes in more rugged hardware form with less processing power and power consumption; this makes it a better choice for some remote, harsh environments.

PAC

Programmable Automation Controller. An advanced form of a PLC, it's equipped with more computer-based capabilities and more powerful resources than a PLC. It can integrate with the Enterprise system and manage control loops. It's more flexible and has multi-domain functionalities.

In summary, these three devices (PLC/RTU/PAC) are similar in role and function; from now on, anything that applies to PLC will be considered as appropriate to RTU or PAC unless it is stated otherwise, but it will be easier to refer to one device during this book, where when we discuss the electric power industry, we will use **IED**.

IED

Intelligent Electronic Device. Hopefully, you won't get confused. IED is similar to PLC in essence, but it's used for the electric power industry. It's a controller based on a microprocessor; it can issue control commands such as tripping circuit breakers if they sense voltage, current, or frequency anomalies. It can also raise/lower tap positions to maintain the desired voltage level.

It comes in different types, such as:

- Protection relay protects power lines, generators, motors, and transformers.
- Bay controller manages voltage regulators, logic in circuit breaker, and event recording.
- Merging units or metering devices take milliseconds, control the circuit breaker to close or open for operation.

Usually, contain around 5–12 protection functions and 5–8 control functions controlling separate devices.

Some IEDs are designed to support the IEC61850 standard for substation automation, covered later in this book.

Others

You must expect many devices that are similar in function. Still, different in capacity and capabilities, for example, as we have PAC and PLC, we also have the Industrial PC (**IPC**), another form of rugged computing control that can serve any purpose and has better capabilities and capacity than PLC and PAC. You may realize now that different devices fit as per the requirements. Look at the following comparison:

	PLC	**PAC**	**IPC**
Processor	Specific purpose	General purpose	Multicore
Protocols	Limited	Wider coverage	Covers all
Programming	LD, FBD, ST, IL, SFC	Support more languages	Support all control languages and computer languages (ex: C++)
Memory	Low	Medium	High
Logging Capabilities	Low	Medium	High

In essence, the controller will have processing power that fits the needs; it can be programmed to control and monitor the I/O modules; you can expect a variety of device types, from Cybersecurity view we see them as control that is communicating over an Ethernet network and are connected directly to field devices.

PLC Programming Language

There are five PLC programming languages in the IEC 61131-3 Standard, which can be used in DCS and SCADA controllers, but it's common to be called PLC programming languages:

1. Instruction List (IL):
    - In the form of a list of textual instructions.
    - It's a low-level language.
    - Very similar to the Computer Assembly programming language.
    - Also referred to as Mnemonic Programming Code.

- Each instruction is in a new line.
- The instruction will have components as follows:
    - Label
    - Operator
    - Operand
    - Comment
- Let's take an example:

We have a switch (On/Off), let's call it (SW1), and a motor that can also be (On/Off) let's call it (MO1), if we want to connect them to a PLC, SW1 will be Input and MO1 will be Output.

If we want to power the motor as On/Off as per the switch state On/Off, we will write the following two instructions:

LD SW1 ← *This instruction will load the state of SW1 and store it in the memory*

ST MO1 ← *This instruction will store the value from memory to MO1*

2. Structured Text (ST):
    - It is a text-based programming language.
    - It is powerful and easier to learn since it's similar to computer programming languages and uses very similar operators and formats.
    - As in the previous example, the ST format will be:

IF SW1 THEN ← *will check the logical value of SW1 is True*
MO1 = TRUE; ← *If the statement is true, set the value MO1 to true*
END_IF;

- If you are familiar with computer programming languages, you will find ST is very similar and easier to learn.

3. Sequential Functional Charts (SFC):
    - Graphical programming language.
    - It simulates the process control flow in graphical charts.
    - It has several elements that can be described:

- What controls are to be executed (steps)?
- When they should be executed (transitions)?
- How they are going to be executed (actions)?

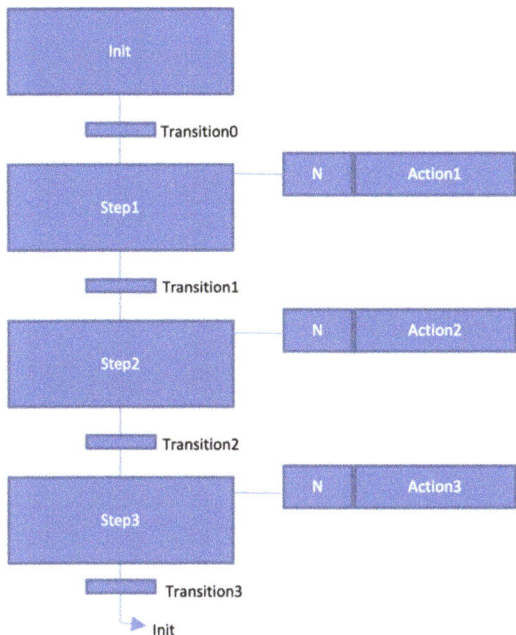

4. Functional Block Diagram:
   - Graphical-based programming language.
   - Illustrates signals and data flows through graphical blocks.
   - The function will be illustrated in a box that receives input from the left side and provides output to the right side.

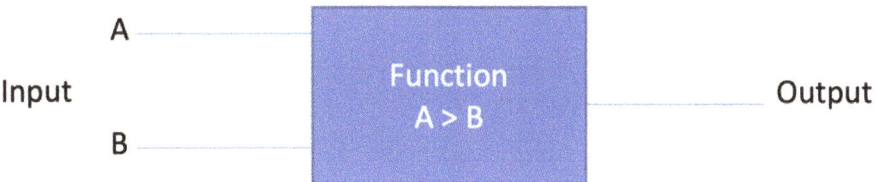

5. Ladder Logic Diagram:
   - The most common language of PLC programming.
   - Graphical-based programming language.

- It's in the form of a ladder, with two vertical lines (power rails) and many horizontal lines (rungs) connected between the power rails.
- The program is read from top to bottom and from left to right.
- The rungs have input/output and logical functions.

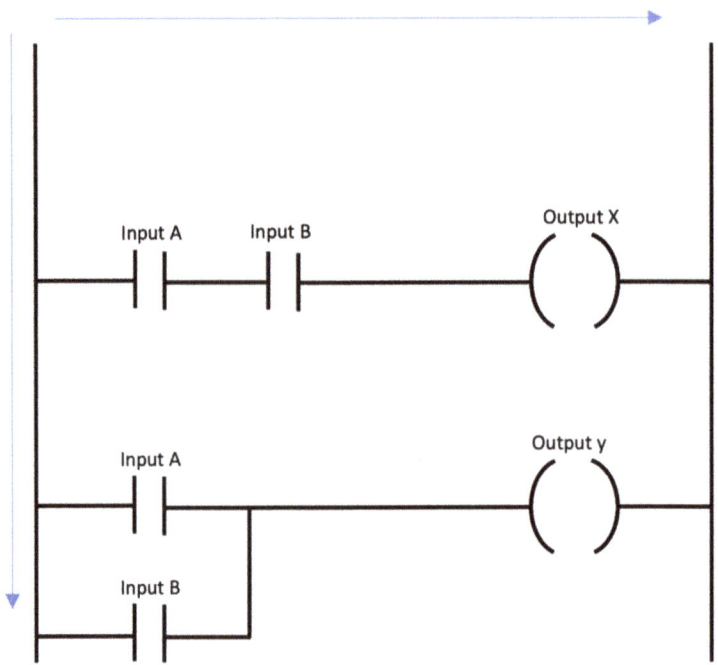

6 The LD logic above is read as follows:
   o   If Input A AND Input B are true, then Output X
   o   If Input A OR Input B is true, then Output Y

Alarms

SCADA is about the acquisition of data and detecting errors; when an error is detected, it will alarm or alert depending on the severity of the detected error.

Alarms mean conditions are met, and action must be taken, unlike events, which don't always require attention.

Alarms have different severity levels (Critical, High, Medium, Low, Diagnostic); you may find Critical/High alarms integrated into HMI to guarantee the operator's attention.

Part of the system design is to define the critical alarms and configure them in the system by either the OEM vendor or the system integrator. They will also need to fine-tune the alarms to avoid duplicates of known issues.

Alarms can be categorized and handled accordingly by concerned departments, and during maintenance, alarms will be ignored.

Alarms can be programmed to take actions, such as email, SMS, activation of a procedure, or trigger other alarms.

Project

You have obtained a good level of what the OT world's components look like; you have an idea about how the logic of automation works; I invite you to have a practical experience.

You may buy a PLC and follow some courses available for free on YouTube or on low-cost training platforms about doing simple projects by programming a PLC to make simple automation.

I have found that playing with the Raspberry PI (microprocessor computer) or Arduino (microcontroller board) is more cost-effective and IT-friendly.

Raspberry PI is cheap, and it has an I/O interface (GPIO) that can be programmed by any language; also, some open-source projects can turn it into a virtual PLC; the trick will be understanding some basic electronics and the concept of relays.

Python programming has a library ready to use to interact with the GPIO of the Raspberry PI, and you can buy low-cost sensors that provide the readings and can use relays for actuators.

Output Relays

They are the output devices that can turn computer logic into different voltage levels; if your device can send an output signal of 5v and you may need a 48v DC signal, you will use output relays that can do this function.

Usually, the output signals from controls go through relays and then connect to field devices at the appropriate voltage level.

Chapter 5: Operational Technology (OT)

You reached the point where SCADA and DCS are classified under the Operational Technology umbrella, in an earlier discussion we referred to various examples of industries, some were industrial (manufacturing) and others were for management of a physical system (building management).

Despite the differences in nature between these systems, they are common in the part that they are physical systems monitored and controlled by cyber systems, and that's somehow the definition of operational technology.

Operational Technology (OT) is hardware and software that detects or causes a change through the direct monitoring and control of industrial equipment, assets, processes, and events.

Operational technology also includes any cyber-physical system and not necessarily industrial in nature, such as medical systems or building management systems.

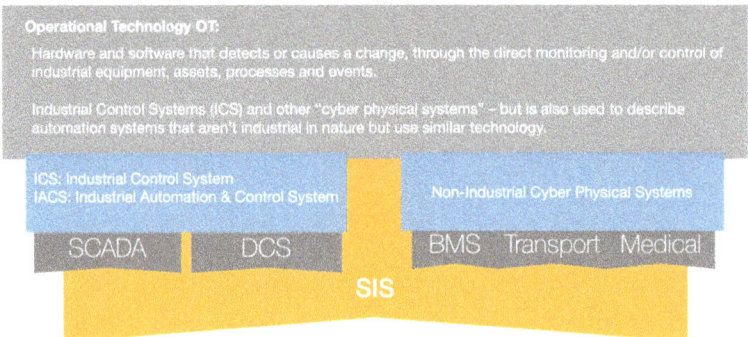

You may notice in the above illustration, the Safety Instrumented System (**SIS**) is shared with all because *safety* is a top priority for all OT systems.

OT is an extensive term; it covers all verticals and industries as long as cyber systems manage physical systems.

DCS and SCADA are classified as ICS or IACS. Industrial Control Systems (ICS) and Industrial Automation and Control Systems (IACS), the terminology in OT might be confusing because, in the past years, the OT term wasn't familiar. Most often, the researches and articles were referring to ICS Security, some were only speaking about SCADA, and only recently they started to use the term OT that defines all Cyber-Physical systems; this point is important because it may confuse you while reading from other sources. You need to know that, in essence, they are speaking about the same thing.

Those from IT backgrounds might get confused because different terms might refer to the same things in OT, and this book should help you understand and align terms to the right meaning.

OT includes many IT components within, but the IT components serve different purposes from the usual IT use in the home and office. You will find the same known OS, PCs, servers, network switches, routers, and even cybersecurity controls like Antimalware, EPP, NGFW, and many others.

In IT cybersecurity risk management, the CIA triad is well known as the top priority goal; you need to maintain confidentiality, integrity, and availability of the systems.

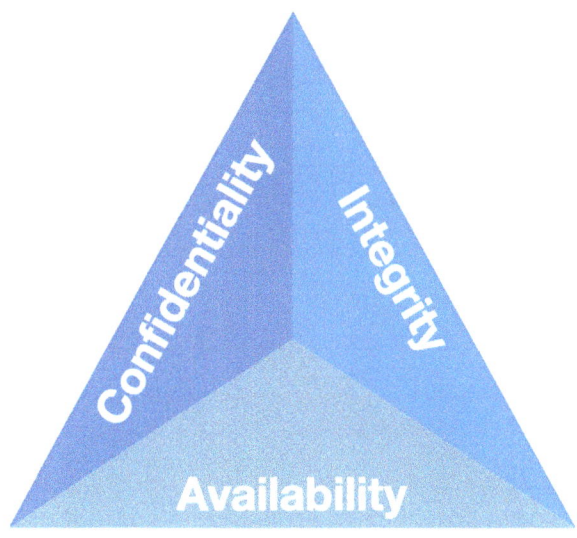

But in the case of OT risk management, *safety* is the top priority, followed by integrity and availability, where confidentiality is still essential but not as it's the case in IT.

Let's discuss some differences between IT and OT:

Comparison	IT	OT
Objective	Managing data	Managing physical systems
Top Priority	CIA	Safety, Integrity, and Availability
Encryption	At rest and in transit is very critical	Some data at rest may be required
Life span	Average of five years	Average of 20–40 years
Availability	Downtime may have an impact	Downtime is critical and will have impact
Maintenance	Regularly	Scheduled in small windows of time
Patching	Applied as recommended	After evaluation, tested, and approved
Cybersecurity awareness	Good	Generally poor in the field

internet Access	Normal	Prohibited and must be isolated
Delays	Delays are accepted	Delays not accepted
Throughput	Requires high bandwidth	Low bandwidth
Operating environment	Controlled	Harsh
Vulnerability	Active scanning	Passive scanning
Data Loss	Backup and Restore	Critical
Availability	99.99% is the objective	100% is the objective

From the table above, you can understand that there are core differences, but what is crucial for you to know is that the most critical difference here is the *Impact*.

Impact of IT system corruption and downtime due to human error or cyberattack or other reasons will usually scale from shallow effect to reputation and loss of $$. It all depends on what type of business; if we are discussing a university website, it will be different from an online store and an airline ticketing system. Some attacks resulted in six-figures fines, and people lost their jobs.

But what is the impact in case the same happens to an OT system? To get the complete picture, I will refer to general theories called the "Butterfly effect" and the "Domino effect."

The flapping of a butterfly's wings can cause a hurricane on another side of the world. You ever wondered how something so small could be a reason for chaos.

A domino effect or chain reaction is the cumulative effect produced when one event sets off a chain of similar events.

In OT, a single small event could lead to an explosion and then a chain of explosions; we want to avoid that at any cost.

Suppose an attacker hijacks readings from a sensor and overwrites the value to a different one. In that case, it could lead to a disaster, and we're not speaking about the loss of money and reputation only, but about casualties!

Our social life order today depends on the resilience of OT against cyberattacks; as per a recently published paper from Gartner (July 2021), they predict that by 2025, cyberattacks on OT will be weaponized to successfully harm or kill humans!

Early in the first chapter, I shed some light on this aspect. Still, it's imperative after you have enough idea about the OT systems to understand how critical your mission is as an OT Cybersecurity Professional.

If a Cyberattack was successfully launched against electrical power generation or distribution, can you imagine the impact on our social life order?

A cut in power impacts hospitals, fuel, buildings, and food supply. Unfortunately, this happened before where attackers managed to control all HMI's power systems and left a city in the dark for almost a day!

Which is more secure? IT or OT?

Security is always the enemy of usability. In IT, all ideas were turned into a working application at first; then they realized they needed to secure it; the Internet today is still running on some non-secure protocols.

IP protocol is not secure, then there was the introduction of IP Security (IPSec VPN) later, but it does not apply to all types of applications. DNS is another non-secure protocol running over UDP. Protocols like HTTP, SMTP, POP3, FTP, and many others were secured later by encrypting them using SSL, so security comes after is a common practice in our world.

Now and then, we hear about severe vulnerability in some OS or Apps, the SSL Heartbleed bug, or as I am writing these words, we're facing the Log4j exposure, it puts all those in the cybersecurity world in panic mode, trying to identify the impacted system, finding good patches, and applying them urgently.

Please don't get excited about OT because, although IT is very risky and vulnerable, it's rapidly fixing its cybersecurity problems. Also, almost every possible attack on IT is possible on OT!

In IT, authentication with complex passwords is a must for almost all sorts of access; disaster recovery and high availability applications make it possible to recover from cyberattack, including ransomware, quickly.

Most OT components don't require authentication, and they are insecure by design! If an attacker managed to reach PLC over the IP network, they most probably could have complete access and control.

In OT, it's not constantly patching vulnerabilities and fixing the bugs; penetration testing is not performed because a simple port scan to a PLC may crash it!

In OT, many plant owners are unaware of how many assets they have and what type of assets to expect, which makes it almost impossible to secure what you don't see or detect.

IT is more exposed to cyber-attacks because it's connected to the Internet almost all the time. Users can use personal computers and other BYODs and expose the corporate network to considerable risk, whereas OT is usually air-gapped and isolated.

Although OT is weak if assessed by cybersecurity posture, it's less exposed to external cyberattacks, where IT is at a better cybersecurity posture but more exposed to external cyberattacks.

So I can't determine the answer to which one is more secure, as we know each operation and each corporate are unique and different in what types of systems are in use and what risk management and security controls are implemented. Generally, I can't confirm that the OT network is air-gapped or that the IT network is exposed.

Air-Gapped OT Network

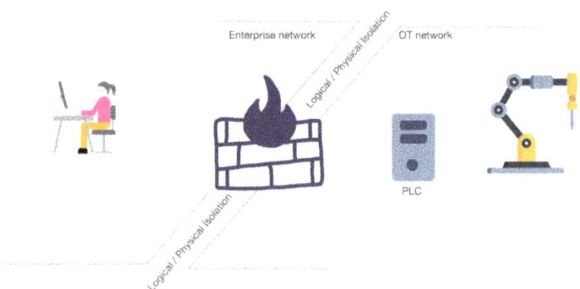

Some incidents that happened in the past and are unfortunately still happening today proved that there's no such thing as 100% isolation of OT network from the enterprise network or Internet.

On the other hand, when you have a machine connected to the enterprise network and then connected to the OT network, this by itself means we have created an indirect link to the internet! Similarly, when a USB flash drive was used in IT and then in OT, it's an indirect link!

In the shadow of the COVID-19 pandemic and travel restrictions, we witnessed the urge to have remote access to operations for maintenance and support when responsible engineers were not able to be physically onsite!

So what I am trying to say is, the Air-Gap is an excellent security control but not a complete one; we can't assume we have 100% isolation.

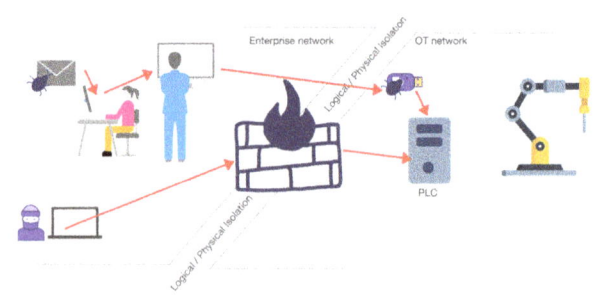

Did you know that phishing is one of the threats to OT? With the assumption that OT is isolated, there are many findings of ransomware attacks on OT networks, initiated by a phishing emails.

We will discuss more in detail in later chapters how to mitigate such risks!

So what has been realized recently by OT organizations and governments is that OT must be reevaluated and observed differently. Instead of being in denial about the possible links between IT/OT along with the technology push toward cloud intelligence to be adapted in the OT world, there has been a real need for accepting the reality and embracing a good plan for creating secure IT/OT convergence.

IT/OT Convergence

Industry 4.0 is the main motive toward the IT/OT convergence, and it's simply by the means of having integration between IT and OT.

To achieve the convergence between IT and OT, it has to begin with people from both sides acquiring exchangeable skills and knowledge; the IT team would have assumptions that won't work for the OT, so they have to be educated on what systems they are going to integrate with from the OT side. The responsibilities also have to be identified between different teams. The OT Team must also be trained on the new technologies they will have to deal with from the IT world.

Once you have both teams trained, they can work together on how the integration will look. All use cases must be reviewed, including the cybersecurity aspects, where you can play a significant role in planning secure IT/OT convergence.

Evaluate the technology in hand, if it's capable of having this type of secure integration or not. Are there any need to upgrade the infrastructure or purchase new software licenses? From the system readiness point of view, it has to be evaluated.

These are general tips on achieving that, but Industry 4.0 also introduces the need for the **Industrial Internet of Things IIoT**, which will be field devices such as sensors providing the readings directly to the cloud, imaging how the change of the attack surface will be shifted!

The OEM vendors will adapt to the 5G network; they will offer field devices and controls that can connect directly to the 5G network, this doesn't only create a link, but it will bypass all existing lines of defense in the network.

In the next chapter, we will discuss the OT infrastructure. Then we will have a dedicated chapter for IoT and IIoT because the OT world today is used to some known references for how an OT network looks. Still, it will take them some time to understand the shift of architecture and attack surface when fully adapting to Industry 4.0.

Part 2: Architecture and Communications

Chapter 6: OT Network Architecture

In all the previous chapters, I tried to let you experience how it feels to deal with the OT world through introductions to new terms and concepts, especially if you are not from an OT background. That was your entry ticket to the OT world.

Of course, engineers from the fields, the actual warriors, are dealing with a tangible world of critical operations in their daily routine. They know what safety is, are constantly trained, and follow many restricted procedures!

They may have a different view of how it feels like to be in the OT world, but you are not prepared to be an automation engineer. Although it's cool, you have a different mission, and you will make sure those in the field of all types of work must be protected from the impact of a cyberattack.

We will begin to get into details of the Cyber part of OT from this chapter, and I need your full attention because it matters!

Purdue Model

In the early 1990s, Purdue University introduced the Purdue Enterprise Reference Architecture (PERA), also known as the Purdue Model, which defines the interconnections between IT and OT, classifying and assigning the assets into six levels.

It considers the nature of each asset in terms of role and function, then assigns similar assets to the same level.

The Purdue model is a logical distribution. The physical systems may be on a slightly different architecture, but it's the best model to visualize the overall operation's assets and interconnectivity.

The following diagram is very common, and it shows an example of how the Purdue model looks like:

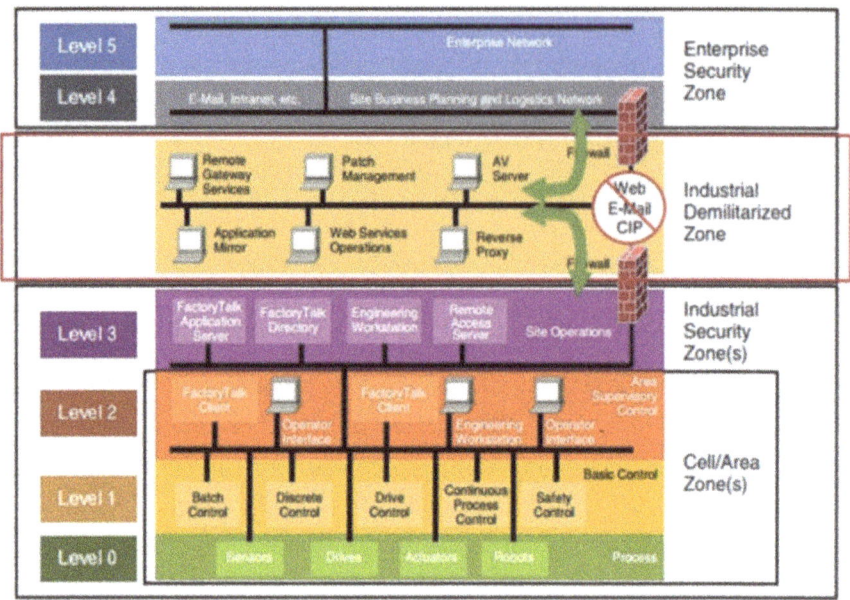

Despite the debate that is taking place today about whether the Purdue model, after almost 30 years, is dead or it still can be used as the primary reference, you have to realize that it's still the only reference for the majority of running operations today. For many, there is no alternative reference.

The main challenge of the Purdue model is Industry 4.0 and the introduction of Cloud and IoT/IIoT that have direct access to the cloud without the need to go through different levels, which changes the architecture entirely, today there are some newly proposed models to adapt Industry 4.0 architecture. But for me, I am confident that the Purdue model is yet to be handy for the next decade at least.

The Purdue model is an easy way to represent the cyber part of an OT; take the following Yokogawa Centum VP digital communication diagram as an example:

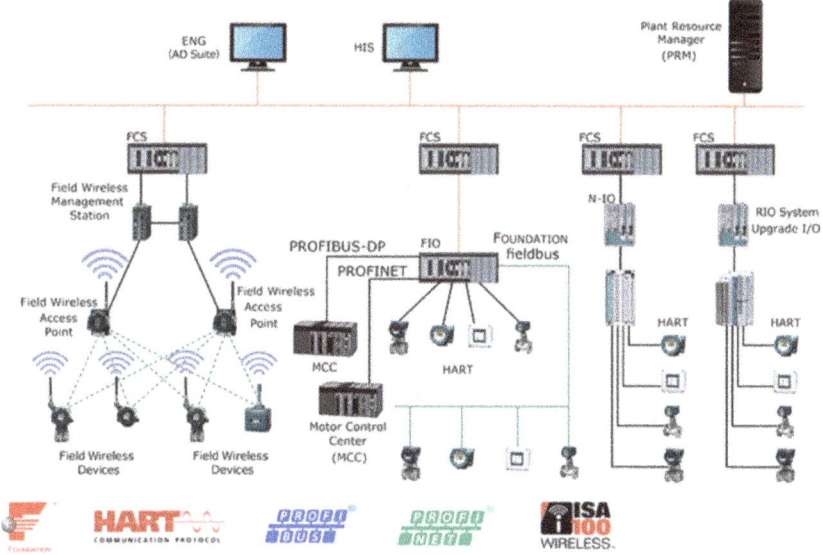

You can directly see the alignment per the Purdue model to levels 0,1, and 2.

In level 0, we have the sensors, some directly terminated to controllers I/O modules and other sensors connected to controllers through wireless.

In level 1, we have the controllers (Field Control Station FCS).

In level 2, we have the supervisory and HMI.

What does each level of the Purdue model represent?

Level 0: Field Devices

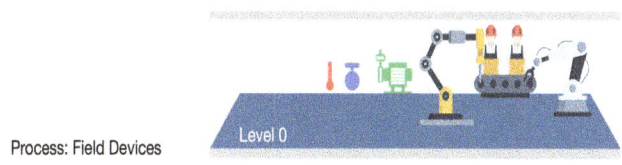

Process: Field Devices

Level 0 represents all field devices in the mean of basic sensors and actuators, intelligent sensors/actuators that can communicate to Ethernet network, and field instrumentations.

Level 1: Control

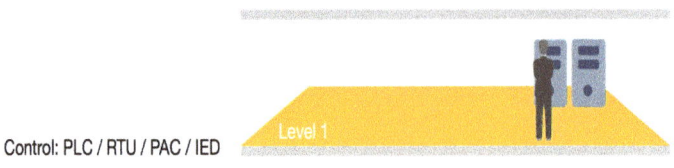

Level 1 represents the control that manages field devices; they can be directly connected to field devices through I/O modules or wireless. Most of the communication with field devices is direct 1-to-1.

Level 2: Supervisory

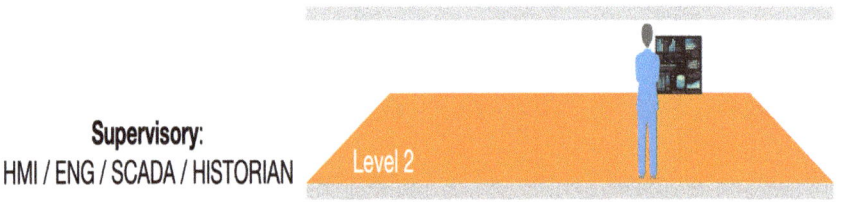

In level 2, we have all the devices responsible for managing and monitoring the control devices; HMIs, Engineering Workstation, SCADA Server, and Historian.

Level 3: Site Operation and Planning

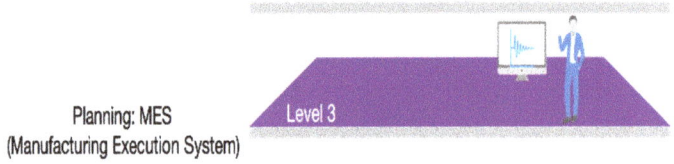

In level 3, we have operation-wide supervisory; the most crucial element is the **MES**, which provides complete visibility to management over the operation processes from start to end. It helps in optimizing production and other decisions (Quality/Quantity).

Level 4: Business Network

In level 4, it's a site IT business network. You can find File Servers, local Active Directory, and most importantly, Enterprise Resource Planning **ERP**, where it manages the production and inventory part of the operation. It's different from MES, where MES is focused on operation optimization. ERP is managing work orders and stock counts.

Sometimes there might be Internet access in level 4.

Level 5: Enterprise Network

It's the IT network as we know it and has the typical enterprise applications and Internet access.

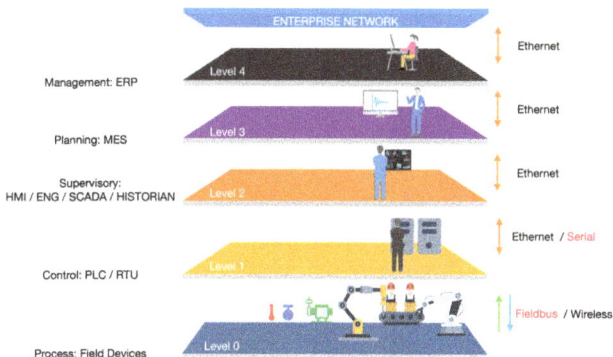

Industrial DMZ (Level 3.5 or 2.5):

The DMZ is meant to create a layer between the enterprise and the Process Control Network; it's not considered a layer by itself. However, it's placed as Level 3.5 between Supervisory (Level 3) and Business (Level 4).

The DMZ has all supporting services and an entry point to the lower levels; you can find an instance of Active Directory and DNS serving the lower levels, Endpoint Protection server, Access Server (Jump Server), which offers sort of controlled remote access to the lower levels. You may find other cybersecurity or IT services that usually serve the lower levels. Essentially, it's a DMZ created by Firewall, which controls the flows and restricts upper-level access.

In SCADA, for example, in power substations, it will contain process control level and field instruments where the supervisory and control are located in a remote location, which creates the need to have a DMZ at level 2.5.

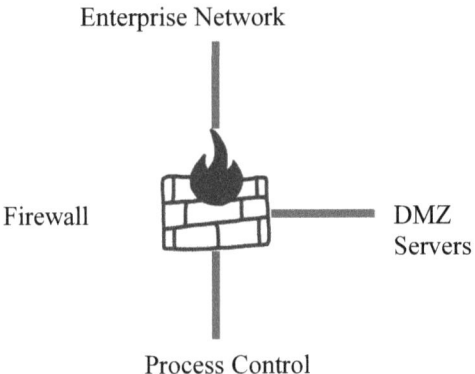

We will speak more about firewalls in the OT Cybersecurity Controls chapter; now, it's essential to know that, in reality, the Purdue model doesn't reflect how the assets are physically connected. You must understand how the assets are classified per role and function. Accordingly, you can identify which logical level of the Purdue model they belong to.

When you start working on some OT architectures, you might not identify the levels and assets' classification right away. Depending on the documentation provided to you and some might be clear; some might be drawings by the engineering software of Schematic Digital Communications. So it's essential to do your homework at this stage. You will need to study architecture more than once. Your goal is to generate a summary excel sheet that includes all assets, roles, functions, Purdue level, and other possible details. You can review the inventory sheet with the operation team to confirm that the details are accurate when you create the inventory sheet.

You need to expect that the OEM vendors' designs will be different. They sometimes have other logic due to various industries and the needs of applications. However, after studying architecture and generating the inventory sheet, you will find them very similar.

Remember, your role is not to understand exactly how the operation works and all the details around each asset. In some cases, when you get confused about an asset classification and what type of communications it offers, you might need to look it up; Google is your friend if the operation team couldn't help, or if you have a contact with the OEM vendor, you can ask them.

Asset Name	Role / Function	Purude Level	OEM Vendor	OS/ Firmware	Application	IP address	Mac Address	Server Protocols	Client Protocols
Plc-10	PLC	1	XXX	X.XX.XX		X.X.X.X Y.Y.Y.Y	11:22:33:44:55:66 11:22:33:44:55:67	Modbus	
Hmi-1	HMI	2	XXX	Win7 / xx.xx	HMI	X.X.X.Z Y.Y.Y.Z	11:22:33:44:55:68 11:22:33:44:55:69		Modbus

Your inventory sheet will look like the above one but might have different details. You may have noticed that the assets have two IP addresses and two MAC addresses. This is normal, for the sake of redundancy, you can find two LANs in the same operation, and each asset has two network interfaces to connect to the two redundant networks, which is reachable on both of them. In case one network fails, the process will continue to run normally on the other LAN.

Another point you may have noticed is the server/client protocol. We will see in the industrial protocols communications chapters how most of these protocols communicate in a Master-Slave way, which in IT is called Client-Server. It's normal to see the HMI Client protocol Modbus, which initiates the connection toward the PLC, which also has Modbus but as a Server protocol; this summary will help specify the Access Control rules that allow/block network communications.

Except for the communications between control and field devices, all other means of communication between the other components use Ethernet. The physical media could be at different Ethernet speeds (10Mbps/100Mbps/1Gbps), and there is wide use of Fiber Optic.

The communication protocols in use can be typical IT protocols (Authentication/web/file transfer/DNS) or OT protocols, which we will cover later.

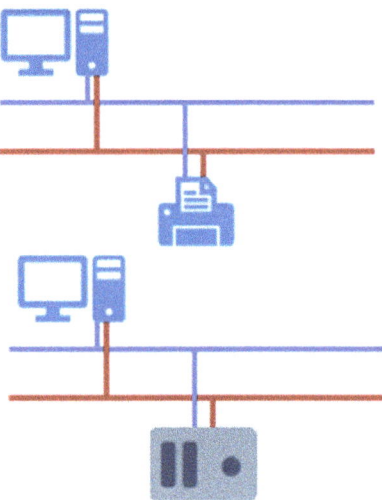

You may find in some diagrams like the above drawing, the two lines (sometimes different colors) represent two LANs for redundancy. And they don't mean it's a linear bus topology; it's for the simple representation of assets. However, each asset is connected to a network switch's port. And for the above example, you can conclude that there are four network switches, where each asset has two network interfaces connected to two separate LANs.

While having these tips in mind will help you read the diagrams that represent the OT Architecture more easily.

Let's add more tips to what you need to know and summarize from the OT Architecture diagram.

Out-of-Band Management Network (OOB)

The network of operation should not be used for the network and security administration. This is not a new concept, usually in IT Network Operations Center (NOC) and Security Operations Center (SOC). The use of a separate physical and secure network for the administration of networking equipment and security controls is a norm.

New networking equipment and security controls have separate physical network cards for OOB management; they are isolated from the operational network.

They don't often appear on drawings and diagrams, but it's essential to confirm the existence of a separate management network.

In case of network issues resulting from cyberattacks or other reasons, the cybersecurity operator must gain access to the network from a separate network that is not impacted by anything taking place on the operational network.

Also, another critical point is that the separate OOB management network can have links to the enterprise network and might offer secure remote access over the Internet. For that reason, we also need to make sure access is restricted only to the management interface of the network and cybersecurity equipment.

Read the Legend

It's common sense, but it's essential to distinguish the different types of physical media in place. The diagram will show connected lines in different colors or dotted lines; each one will represent a purpose, network, or physical media type.

You may see FO (Fiber Optic), and you need to propose a firewall or mirror the traffic to an IDS (Intrusion Detection System). You need to provide a suitable model that can fit the physical media requirements and has network interfaces for Single/Multi/ SFP/ SFP+.

DATA DIODE

They are a one-way firewall that allows communications to go in one direction only; for example, it won't enable TCP communications because it has a 3-way handshake and, 2-way transmission/acknowledgment, and a retransmission mechanism.

DATA DIODE will allow data to be sent from low levels (process control) to the enterprise network, where information must be provided to upper levels for analysis. At the same time, it's important not to allow a single frame/packet to be sent down to the lower levels.

They can be hardware circuit-based or software-based; either way, they provide the same purpose.

In the case of an IDS that needs to receive mirrored traffic from the operation for passive analysis when you have a DATA DIODE presented, it must be considered as DATA DIODES will allow the flow of mirrored traffic through it, which means if the IDS Sensor is placed on top of the DATA DIODE, it will work fine because the mirrored traffic is one-way. If the sensor is below the

DATA DIODE, it needs to communicate with a management console; usually, it will be over TCP/SSL, which is not possible through DATA DIODES, so this must be considered in the design and proposal. However, if there's an OOB management network, this won't be an issue. We will discuss it in more detail in the cybersecurity controls chapter.

Traffic Mirroring (SPAN)

Passive cybersecurity techniques are highly desirable in the OT world because most of the assets within OT are sensitive. Any means of active scanning on the network most probably will cause issues; I know for a fact that if you perform a massive active network scanning to a critical operation, you will crash it and bring the network down, so never do massive active Pentest on OT, only controlled supervised scans if required.

For that reason, there are cybersecurity controls that rely on a copy of the traffic and not the actual traffic, hence the need for the network switch feature of traffic mirroring or SPAN (Switch Port Analyzer).

The SPAN is a dedicated port on a network switch that receives mirrored traffic from one or more ports/VLANs; the SPAN port will be connected to a monitoring device that has the intelligence to inspect for specific criteria.

Some SPAN can include Layer 2, and above of the OSI Model (Data Link Layer), which consists of the Ethernet Frames, and others only possible to provide Layer 3 SPAN, which includes only IP Packets, that means if the monitoring device is looking for Ethernet Frames for inspection or the operation uses non-IP protocols. They are the different means of Ethernet Frames protocols, then this won't work; usually, in this case, it might be due to the presence of IPsec VPN tunnels and the need for SPAN to a remote location. IPsec VPN tunnels only offer Layer 3 (IP Packets). However, there is some possibility to create Layer 2 GRE tunnels inside IPsec, which can be used for Layer 2 SPAN. This is too complex a configuration, and it's not recommended. You need to know the SPAN requirements to check the available options within the operation and advise your recommendations to have a complete Layer 2 SPAN if mandatory for the monitoring device.

If you find switches that don't support SPAN, check if there are other points in the network that you can mirror. To make a bit more precise, you may have L2 access switches that don't support SPAN and are connected to distribution switches that support SPAN; you need to check if there is a possibility to SPAN the access switches' traffic from distribution switches. If that is the case, then it does the trick. Other than that, you need to advise to upgrade the network switches to new models that can support SPAN or you may propose to use a network **TAP** device.

You also need to take into consideration the traffic volume and capacity of switches; the SPAN adds more load to the switch processor; if the load is already too high, then enabling SPAN might cause issues that we don't need, so bear in mind the capacity of the switch versus available resources.

Finally, you will need to build a matrix as part of the inventory that lists all network switches, made, model, number of available ports, number of free ports, and the support for SPAN.

Network TAP

TAP (Traffic Access Point) is hardware that can mirror traffic when SPAN is impossible. If you have a link between two network assets (A & B) and you need to analyze the communication between them, you can connect both assets to the TAP ports. It acts as a passive connector, while the TAP will provide a copy of the traffic through a third port.

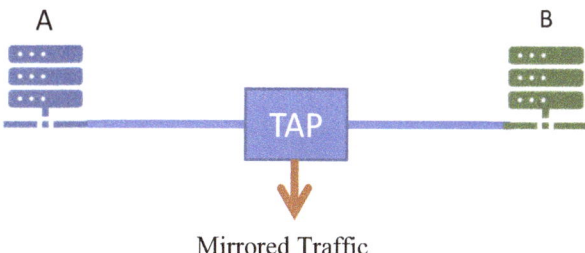

Mirrored Traffic

TAP devices come in different models/ports configurations, and interface types; when required, you need to find the proper model to fit the operation.

Virtualization

You may come across virtualized systems; some OEM vendors provide applications in virtual machines; virtual machines are assets that must be considered and added to inventory along with the Hypervisor Host(s).

A virtualized environment has the same network functionalities as a physical one; you can find all means of routing and switching. Cybersecurity controls are available as virtual machines deployed within the virtual environment itself.

If you need to mirror traffic from virtual machines, you need to check the hypervisor capabilities. Within the virtual switches, you must enable the security options that allow mirroring. You need to enable the promiscuous mode to enable the monitoring device to receive all traffic.

Suppose the monitoring device is virtual, and you need to mirror traffic to it from the physical network. In that case, you will need to use a dedicated network interface on the host and enable the passthrough (assigns hardware directly to a virtual machine without going through the hypervisor network driver).

NAT

Network Address Translation (NAT) can alter the source/destination of an IP / Port; NAT is usually helpful in networks that have access to the internet to translate all private IP addresses into one or more public IP Address(es), also can be used in publishing services through Public IP.

Usually, NAT is associated with **PAT** (Port Address Translation), which is used to change the source port to avoid the possible duplication of connections and for security reasons to avoid the possibility of fingerprinting the OS version by knowing which source port is in use.

It's not recommended in OT, especially at low process control levels; for monitoring devices, it will be challenging to distinguish assets since the IP address is hidden due to NAT, same as the source port number can also be hidden.

Similarly, L3 Routing can hide the MAC address of the assets, limiting the visibility of assets; we always recommend having Level 3 and below to be a flat network without introducing any routing at these levels.

Throughput

Unlike IT, OT will always have less bandwidth utilization; for the same number of assets, they might have 50X more traffic in IT than if they are in OT. The reason for this is that IT assets will be running many applications, and they establish tens of connections in the background for refresh and regular updates. A smartphone could have 100 opened connections even if you're not using it. In contrast, in OT, the assets are dedicated to working on specific functions and applications.

The transferred data in the IT world is massive considering the different media of voice, video, documents, and other large files, wherein OT always moves tags, which is significantly less than the case in IT.

You need to estimate the OT throughput as part of the inventory you are building; this will play a significant role in selecting the proper cybersecurity controls capacity.

Environmental and Form Factor Requirements

Due to the different operations, they have additional operating environment requirements. You will need to decide on what type of equipment fits.

OT Network diagrams might or might not provide information about the environmental requirements; for example, you may see some drawings of cabinets, which means you need to understand the form factor of equipment supported in these cabinets and available space in the cabinet to add more devices; in other cases, there might not be details about the physical specs of the supported equipment, so you have to obtain this information.

If you can identify what type of infrastructure is in use, it will help you to know how harsh the environment is; if you find a rugged network switch, lookup up the model and check the specifications, and you will have an idea about what type of operational environment and form factor required.

While building the inventory of assets, as explained earlier, you have collected enough information about what types of assets you need to secure; at a later stage, you will need to propose cybersecurity controls to mitigate the cyber risks, now when doing so, you will follow the details provided in later chapters about the cybersecurity controls, but for now, you need to identify the operating environment requirements of all network and cybersecurity equipment.

Rugged Hardware

It's the type of hardware that can operate in harsh environmental conditions, and it applies to many kinds of equipment such as network switches, routers, firewalls, servers, appliances, and PCs. And based on the operation, they have to be compliant with specific operational standards.

Following are essential specifications of the rugged hardware:

1. Operating Temperature

Depending on which part of the world the equipment should handle a broad operating temperature range, you may find rugged servers that support from -50° C to +50° C (-58° F to +122° F). If you are in the Gulf countries, they have recorded 53.2° C in June 2021, which means this range is not enough considering when the hardware is operational it will have a higher temperature than the environment around it, so that means when you see requirements for support up to + 85° C operating temperature is considered normal.

Again, it all depends on the environment of the operation; this is important to identify and make it part of a matrix that you need to build for the compliance of any proposed hardware to fit into the operation.

Don't confuse operating temperature and storage temperature; storage temperature is usually higher than operating temperature, but we're concerned about when the hardware is up and running. What is the maximum temperature it can continue to run normally at?

2. Humidity

It will be presented in a percentage format (example: 0–95%); make sure to add the humidity requirements to the hardware compliance matrix.

3. IP Rating

IP (Ingress Protection) rating as defined by the international standard EN 60529 provides information about the hardware capability of protection level against solids and liquids.

The rating of hardware comes in the form of 'IP' followed by two digits, ex: IP30.

The first digit indicates the level of protection against the ingress of solid objects, whereas the second digit indicates the level of protection against various forms of liquids.

The following table shows the reference chart for the IP Rating:

1st Digit	Solid Protection	2nd Digit	Liquid Protection
0	No protection	0	No protection
1	Protected against solid objects over 50mm, e.g. accidental touch by hands	1	Protected against vertically falling drops of water, e.g. condensation
2	Protected against solid objects over 12mm, e.g. fingers	2	Protected against direct sprays of water up to 15 degrees from the vertical
3	Protected against solid objects over 2.5mm, e.g. tools & wires	3	Protected against direct sprays of water up to 60 degrees from the vertical
4	Protected against solid objects over 1mm, e.g. wires & nails	4	Protected against water splashed from all directions, limited ingress permitted
5	Protected against dust limited ingress, no harmful deposits	5	Protected against low-pressure jets of water from all directions, limited ingress permitted
6	Totally protected against dust	6	Protected against strong jets of water, e.g. on ships deck, limited ingress permitted
		7	Protection against immersion in water between 15cm – 1m deep for 30 minutes
		8	Protection against immersion in water under pressure for long periods
		9K	Protection from close-range powerful, high-temperature water jets

4. DIN-Rail

The rugged hardware can be mounted on DIN Rail; it's different from Rack.

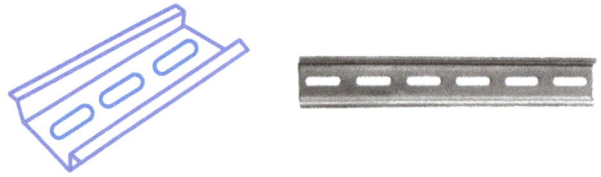

It's a support structure where rugged appliances and electrical components can be mounted.

It may come in different sizes; you need to know if the rugged hardware devices are compatible with the Din Rail (35mm, 32mm, or 15mm).

5. Power

You need to know what power source is available and if redundant sources are available. Power specifications must be met, such as AC/DC, voltage, ampere, and in some operations, it's essential to check the power consumption because there might be limits.

Connector types have different standards in regions of the world, which means you need to know the supported power cord and plug and check if there is a need for a power adapter.

1) Special requirements

In some operations, there will be a need to have hardware that can withstand shock and vibration, or there will be a need to have a special coating for hardware; pay attention to these details as they will be represented in the form of a certification of compliance with the standard.

For example, in power substations, the rugged hardware must comply with some standards such as IEC 61850-3 or IEEE1613. When you check the hardware, it must be compliant with these standards, which is in this example to withstand the ElectroMagnetic Interference (EMI).

What are the network topologies in OT?

You may expect different designs, and here we will discuss the most common examples.

Redundant Star Topology

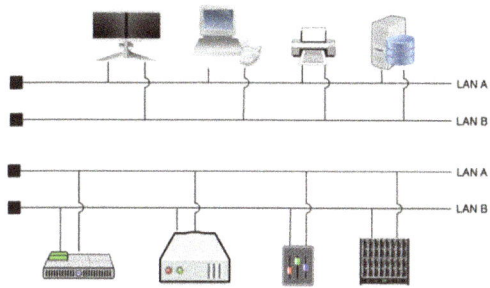

As discussed earlier in this chapter, you may find a linear drawing, and it indicates that it's a star topology, which means all assets are connected to a network switch port; the following diagram will show how they are connected:

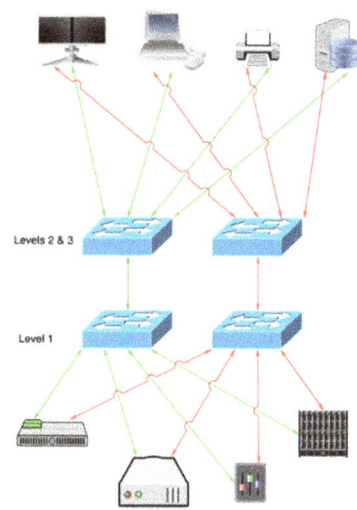

You have Level 1 network switches, and both levels 2 and 3 are connected to the same network switches; for the sake of redundancy, we have two separate networks (LAN A and LAN B), and although during the discussion of the Purdue

Model, we separated between Level 2 and Level 3, you can notice in reality it's different in how the assets are connected.

This also helps you when building the inventory sheet to classify per the Purdue Model following the role and function of the asset, not how it's physically connected. And it is an observation to highlight that both Level 2 and Level 3 are on the same flat network. If there were a need to implement cybersecurity controls between these two levels, it would imply the need for separating them into new network switches.

Having redundant networks requires a sort of network protocol to manage the redundancy, wherein in this case, Parallel Redundancy Protocol (**PRP**) is used. In PRP, each node sends the traffic duplicated in each different LAN, and the receiver consumes the first frame and discards the remaining ones. Not all assets support PRP, and in this case, there is a need for a Redbox device that can connect to the asset on a single port and connect to the redundant LANs using another two ports.

Redundant Ring Topology

In this topology, each asset has two possible paths; if there was a link cut in one path, the asset could communicate through another path; the network switch decides which path to use; assets are not aware of which switching method is in use.

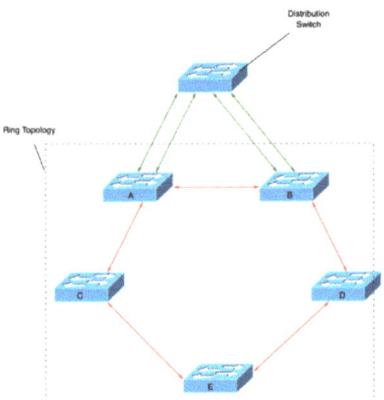

The network switches will be stacked in the cabinets and then connected to the distribution network switch.

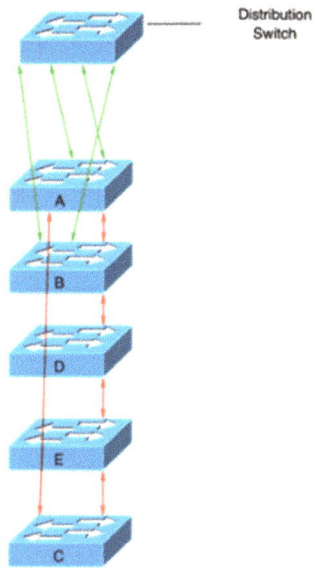

Spanning Tree Protocol (STP) is often used in switching technologies when multipath avoids switch loops. Still, STP is very slow; it can take seconds to network recovery, which is a long duration for critical operations.

Instead, two protocols can be used:

Media Redundancy Protocol (**MRP**) is a protocol used to avoid a single point of failure; the network recovery time using MRP would be 10ms or less.

High Availability Seamless Redundancy (**HSR**) is a protocol that provides seamless failover against the failure of any single point. The traffic is sent in both directions, and in the case of a cut in the link, the traffic is already received through another link and without the need for a recovery mechanism.

You may find multi protocols on the same network, such as HSR and PRP, and when the asset or networking device doesn't support the used protocol, they use a Redbox.

You have to make sure that any proposed cybersecurity solution can fit with the topology and redundancy protocols in use.

In summary, when you study an OT network, you need to understand the role and function and other details of each asset, assign them to levels as per the Purdue Model, summarize the communications flows into a table, identify topologies and protocols in use, build a matrix of networking equipment along with details about physical media in use and SPAN capabilities, confirm the availability of the OOB management network, and list all environmental compliance requirements.

This will be your baseline for any cybersecurity solution that you will propose. It has to fit the criteria you have built.

Chapter 7: IoT and IIoT

"Why should I care?" Sadly, I heard this question during a side conversation about the importance of IoT cybersecurity in gathering back in 2018.

It was two years from October 2016, when the Mirai Botnet Attack took place; Twitter was down for about two hours! It was a massive DDoS attack; some stated that approximately 300K IoT devices were compromised and became part of the botnet that launched the attack.

Maybe we need to ask ourselves, what does it mean to have approximately 30 billion connected devices in 2022? Or how many of these devices are not secure and easy to hack?

It's not about who will hack my smart thermostat. Are they going to play with my room temperature? It's not that bad! Well, it's about if your smart device got compromised; it's just the beginning.

Do you know that you might be held accountable if an IoT device you own is part of a cyberattack! Do you know that IoT devices can compromise your privacy? Do you know that a compromised IoT device can be an entry point to compromise the complete network at home or office?

I am not being dramatic about it; I am trying to reflect the reality! Because bad things happen, and if we are not careful enough, they might happen to us! I only have concerns about the marketing campaigns pushing insecure IoT devices to the consumer market while promoting how efficient and effortless life will be.

Internet of Things (IoT)

Consider any physical object, such as household appliances, office equipment, or any physical device that has software and does specific functionality; when these physical objects get network capabilities, it's called the Internet of Things.

Well, they also call them smart things; now, you can get your smart kettle and innovative coffee machine; if you are too lazy to walk to the kitchen to press the button to make your coffee, you can select your coffee from a smartphone app.

The smart home has various options to control temperature, lights, doors, and CCTV cameras. It's a way of having a luxury lifestyle and is used in offices and industries.

We can consider any device that has network connectivity without a user behind it an IoT device; this includes printers, cameras, digital signage, billboards, traffic lights, drones, and it can be helpful in any industry, medical devices, industrial, municipalities, road, and transport, everywhere we can have IoT.

Other criteria to classify an IoT device are that it's impossible to deploy an agent software; they are shipped with closed-manufacture software.

They are connected to the network through Ethernet, wireless, and 4G/5G; you will also find Bluetooth-enabled devices.

They are not managed in nature as they are supposed to function independently without users' interaction. However, there are monitoring tools, and known cloud-based platforms used to monitor IoT devices and alert in case of cybersecurity events.

Based on the above, we can identify common IoT properties:

- IoT devices function independently of users.
- Not possible to deploy software agents to them.
- They are shipped with manufacture firmware.
- They have a network interface (wireless, wired, 4G/5G).
- They are unmanaged.

Having the above said, why is the market always combining IoT with AI and cloud? You must know that AI is only in the cloud and uses the data it collects from IoT devices. The IoT devices are not smart or intelligent, but they are dummy and programmed to do basic functionalities with networking capabilities.

Industrial Internet of Things (IIoT)

IIoT is a network of physical field devices, sensors/actuators where some have embedded controllers, a subset of the IoT developed and used in operational technology.

We have come across Industry 4.0, where IIoT is the field device to share data with the cloud for analysis.

Cloud is just where we host the hardware/software/application. It has an infrastructure and several data centers in different regions, so we need to understand all the discussions about collecting information from field devices and providing them to the cloud for analysis.

We can refer to the proposed Reference Model for an IoT infrastructure by the *IoT World Forum*, which provides a high-level design and architecture for the application of IoT/IIoT and cloud-based AI. They have proposed seven layers as follows:

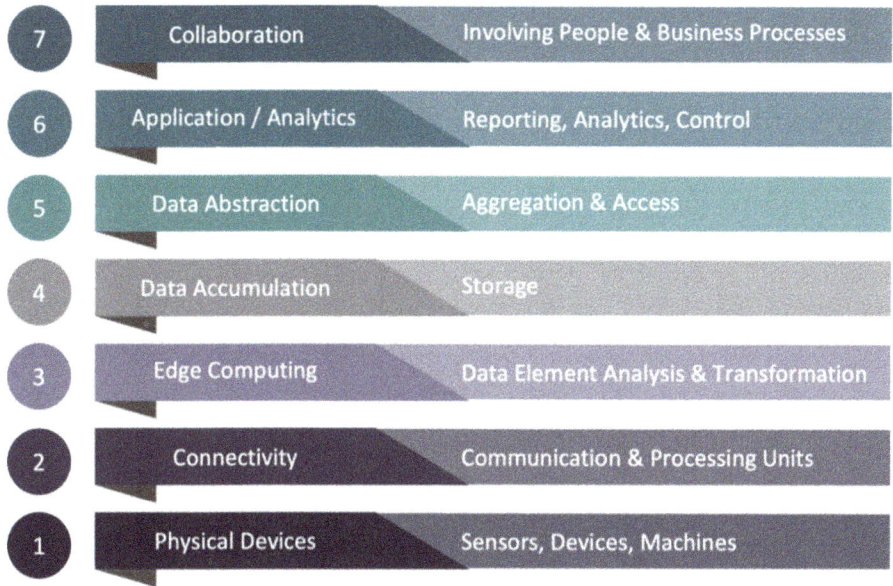

Layer 1: Physical Devices

The IoT/IIoT devices act as sensors or actuators.

Layer 2: Connectivity

The communication media (wireless, wired, 5G, Bluetooth) and communication protocols.

The IoT/ IIoT might be connected through an enterprise network infrastructure and services. They might as well be connected directly through the 5G network to the cloud, eliminating all enterprise perimeter control and cybersecurity.

Layer 3: Edge Computing

This layer performs filtering and normalization of the data received from IoT/IIoT devices to be ready for upload to the cloud where we have the database storage.

Layer 4: Data Accumulation

Data, after being filtered and normalized, is stored in a database.

Layer 5: Data Abstraction

Data in this layer is aggregated and made available for access by **Artificial Intelligence (AI)** applications and reporting.

Layer 6: Application/Analytics

Data is being accessed and analyzed; the AI algorithms will be executed to extract intelligent information that helps decide and report the state of operation.

Layer 7: Collaboration

People and processes will benefit from analytics and reports to act and collaborate.

This reference is data-driven that addresses the abstraction layers through the data lifecycle.

How about cybersecurity? As OT cybersecurity professionals, we are concerned about the physical and connectivity layers, but that doesn't mean other layers of cybersecurity are not essential. Still, it's out of the scope of this book, and when data is in the cloud, we assume they already have cyber risk assessment and cybersecurity controls in place.

IoT/IIoT Cybersecurity Challenges

When we listed the common properties of what defines an IoT device, we listed the points that are also considered a challenge for cybersecurity.

- IoT devices function independently of users, which means we are missing the user attention factor; operators could report suspicious behavior. Still, in the case of IoT, it might be compromised and not be noticed for a long time.
- Not possible to deploy software agents to them, so they are off the radar of cybersecurity controls.
- They are shipped with the manufacturer's firmware running on the embedded OS, which might be obsolete, and several studies showed that they are not hardened and vulnerable to elemental attacks. Through reverse engineering, attackers can find many ways to access the device without any authentication.
- They usually have open, non-secure protocols (MQTT, Modbus, HTTP, Telnet, and FTP).
- Hardcoded credentials can be found in manufacturing manuals and guides.
- They have a network interface (wireless, wired, 4G/5G), so they might be outside the perimeter cybersecurity controls.
- You can find default login and commonly used SNMP strings that are not appropriately configured.
- They are unmanaged; they can be accessed remotely for functionality controls but not for cybersecurity management and **R**ole-**B**ased **A**ccess **C**ontrol **RBAC**.
- They are subject to close-range attacks through Bluetooth or wireless.
- Considered the weakest link in the network, and if compromised, attackers can obtain information about the network, for example, the wireless passphrase.

There are many challenges to be addressed from a cybersecurity perspective, but there's also the architecture and attack surface shift! Some MES systems designed for process optimization will deploy IIoT sensors to provide feeds only about the operation's health; they might be connected over 5G directly to the cloud. This will eliminate everything we know about the Purdue model, and therefore there are debates about if the Purdue Model is still the reference for OT or if it must change.

In essence, it looks like a one-way IIoT sensor, but this is a Two-way communication capable device. When this is the case, we need to consider that perimeter security controls are bypassed, and we have Level 0 sensors connected directly to Level 5 of the Purdue model. Accordingly, mitigating the risk of IIoT devices compromise is different.

For IIoT devices and their communications, we are concerned about:

1. Infrastructure security.
2. Cloud security.
3. The cybersecurity posture of the IIoT device:
   - Hardened OS and applications.
   - Assessed for vulnerabilities and weak/default credentials.
   - Disable insecure protocols use.
   - Disable all means of connectivity except the main link to the cloud (USB, Bluetooth, Wireless, Ethernet)

IoT/IIoT Infrastructure

Let's take the example of the MQTT protocol, a standard messaging protocol designed as a lightweight publish/subscribe messaging transport protocol.

It has three main components:

- IIoT/IoT publishers (Edge)
- MQTT Broker (Cloud)
- Subscribers, client software for end-users, and processes (public)

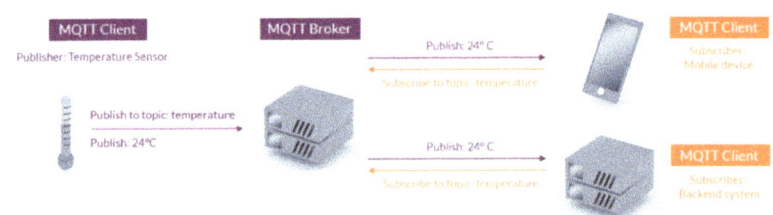

It's an efficient protocol that makes the data available for applications, but we are not discussing the efficiency; instead, we're concerned about security.

What are the common options of network media types for the publisher?

- Wired/Wireless
- 5G

If wired/wireless in use, then it must connect through Firewall, and communication to the cloud must be through a VPN. The challenge initially addressed through IoT is the connectivity and availability of infrastructure. In many cases, the only possible transport media is through 5G, and while 5G is used, as repeatedly highlighted, it has direct connectivity to the cloud.

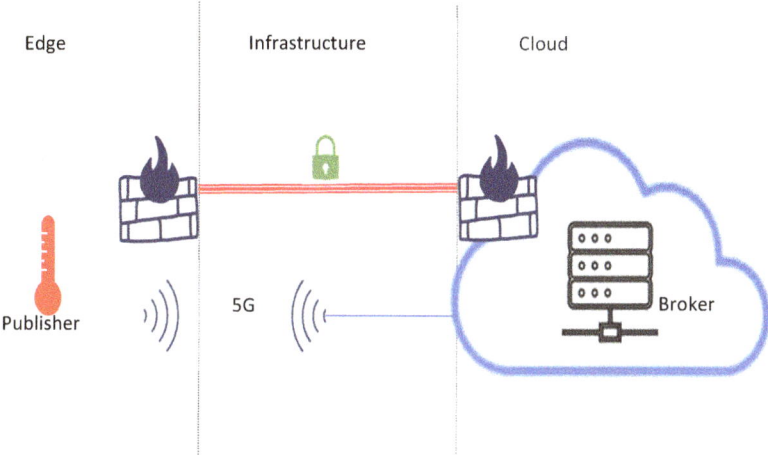

That said, when there's a need for 5G, we need to have the communications encrypted by TLS (MQTTS). In that case, we trust the publisher to have direct and secure access to the cloud, which is the risk you accept in return for achieving secure communications.

The risk of compromised IIoT/IoT devices impacting the cloud will remain if you use TLS encrypted communications over a 5G network, but is this the only risk to be worried about?

5G and Telecom attacks

As for IT/OT, there are attacks specific to telecom networks, and 5G can also be targeted for particular attacks that cybersecurity experts might not be aware of.

The 5G threat landscape includes the possibilities of eavesdropping, interception, and hijacking. Examples of the Telco-specific attacks are:

- Rogue base stations
- SMS brute force
- Fake GTP
- Battery drain
- Bidding down to 2G or 3G

Understanding the risk associated with the 5G network is essential, and even if you are no telco expert, you need to make sure if 5G is used to have a plan for mitigating the 5G risks, consult with an expert in the field and check out the recommendations and best practices to build your mitigation plan.

The finishing piece I would like to highlight is the direct connectivity between IoT/IIoT devices; this would expose more devices to attacks in case of compromise. This aspect must be considered, and the most convenient approach would be the zero-trust and segmentation that we will discuss in a later chapter.

I am always worried about IIoT and IoT devices because they are often off the radar, and they are the blind spots and weakest link in the network; if they are not appropriately secured, you must expect the worst to happen.

Chapter 8: Introduction to OT Network Protocols

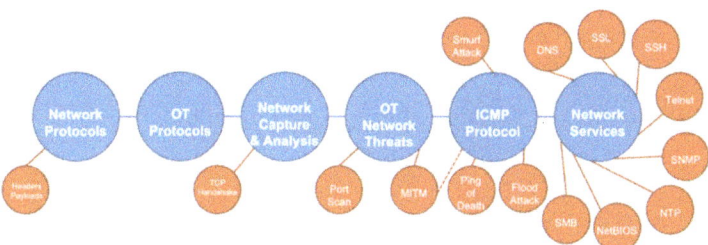

This chapter highlights the fundamental aspects of what you need to know about OT network protocols. Even if you are familiar with these protocols, there are some essential points to be considered as an OT cybersecurity professional. You will explore different protocols in different ways; sometimes, we will analyze the packets; others will list capabilities, where we will always highlight the risk associated with the protocol.

Consider this as a theoretical lab. It will be the practice of exploring the protocols in general, and you can find sample PCAP files to download if you wish to have a deeper look, under the Downloads section of www.otsec.io.

In general, protocols will be used to:

- Exchange data.
- Monitor and manage network devices.
- Security transport layer.

A protocol is a standard reference or language that two different systems can refer to for a unified way of communication. For example, each country has its own rules and system. Still, in the case of agreement with another country about any subject, they can agree on a protocol that can be used to be the rules to follow when these two different countries are working on that subject.

Computers are of different types, and so are the data representations, so when two computers communicate, they don't use their presentation of data. Instead, they have to provide the data in standard presentation (example: ASCII), and they have to use a protocol understood by both to exchange the data following the protocol format.

What will happen if a developer builds a website but uses his specifications on how the data is represented and transferred? No one can browse that website because it must communicate over HTTP protocol; otherwise, nobody can access it because data will be presented in different formats and orders.

For this reason, we have standards that describe the specifications of a protocol, and the Internet Engineering Task Force (IETF) is managing the Request for Comments (RFC) for most known protocols that we use today.

RFCs are in the form of text files; they have detailed specifications about the protocols, and they will explain the format, length, expected values, reserved bytes, etc. It provides a map for how to read a protocol communication.

Some OT protocol specifications are defined by committees or forums or might be proprietary protocols owned by OEM vendors; in the latter case, we don't have public documentation about the protocol specifications.

One robust security control is to have all communications checked to match the standard; this is considered a whitelisting check from the structure of protocol point of view. Suppose an attacker is trying to send malicious traffic by claiming it's a legit protocol. In that case, it will probably not match the standard, which by this mechanism can detect malicious activity on the network even if it's unknown before!

Open Systems Interconnection model (OSI)

When you read an RFC, usually it will refer to the OSI model, which is the first subject taught in networking; I will only focus on essential points and not go through the OSI model details.

It's a seven layers' model that describes the abstract of how the protocol is formed because we use the protocol as the standard format for exchanging data over the network; since data can't be transmitted as a whole, it's split into smaller parts (payload). Eventually, the received data might be a single payload or a collection of payloads. The protocol is responsible for marking and ordering payloads to be reassembled correctly into their original form.

Each protocol has specifications for headers and payload. The header will contain information about the payload, the source and destination, and the protocol in use.

Let's take the Data Link layer 2 as an example:

If you have an Ethernet frame of 1518 bytes in length, we need to read the information from this frame; here is how the bytes are read:

- Six bytes: Destination Mac address
- Six bytes: Source Mac address
- Two bytes: type
- 1500 bytes: Payload
- bytes: CRC for error check. (Tail)
- Header + Tail = 18 Bytes and Payload is 1500 Bytes

Now, this is not the case for all Ethernet frames; for example, you may be using VLAN, and so additional information will be added to the Ethernet frame header as follows:

- Six bytes: Destination Mac address
- Six bytes: Source Mac address
- Four bytes: 802.1Q Header
- Two bytes: type
- 1496 bytes: Payload
- Four bytes: CRC for error check (Tail)

Four bytes were used for the VLAN header, which has been deducted from the payload.

This is an example of an Ethernet frame; if you consider a wireless frame, it will be different, but we use 'frame' as the unit because it's the data of the layer; in Layer 2, "Data Link," the Protocol Data Unit (PDU) is "Frame," check the following table that lists the seven layers along with the PDU:

7. Application	HTTP, FTP, SMTP, IMAP	User Interface	
6. Presentation	JPG, MPEG, GZIP	Compression, Encryption	Software
5. Session	RPC, NetBIOS, PPTP	Session point to point	
4. Transport	TCP / UDP	Segments / Datagrams	
3. Network	IP Address, L3 Switch, Router	Packets	
2. Data Link	MAC Address, L2 Switch, Bridge	Frames	Hardware
1. Physical	Physical Media	Bits	

During the discussion of this book, if we refer to frames, it's the data link layer, packets for the network layer, segments for the transport layer (in case of TCP), datagrams for the transport layer (in case of UDP), payload or data for the upper layers.

Also, we can refer to them differently, such as Layer 3 or IP Layer for the network layer, and it's common to use 'Application Layers' for upper layers 5, 6, and 7.

Layers 1, 2, and 3 are also referred to as hardware, whereas Layers 5,6 and 7 as software; however, the OSI model will become natural for you if it's not the case already.

Why 1518 bytes for an Ethernet frame? And why 1500 bytes for the payload?

These are not standard numbers but most commonly used; if you respect these numbers during transmission, you will avoid many issues. For example, an IP packet typically is 1500 bytes, as in the previous example of an Ethernet frame, when we added the VLAN header it consumes 4 bytes; if we don't deduct that from the payload (IP packet) then the frame size will be 1522 bytes, and the recipient is expecting 1518 bytes.

The IP packet size is configured on systems in terms of Maximum Transmission Unit (MTU), which by default is 1500 bytes. Still, it can be higher; some special applications use Jumbo Frames; on a Gigabit Ethernet network, it can be a maximum of 9000 bytes of frame size.

IP Fragmentation is used when you have two systems configured on different MTUs; for example, one device expects 1400 bytes, and the other is sending 1500 bytes. As a result, the packet will be split into smaller pieces called fragments, and on the recipient side, these fragments are reassembled again.

IP fragmentation, if found in a network, must be investigated. Still, you first need to refer to the standard (RFC 791) to understand how it works and then compare the fragmented packets in your network because some attacks may split payloads into multiple fragments to avoid being detected by cybersecurity controls. And also, the overhead processing for fragmentation and the extra headers added to fragments will have an impact on the performance, so from both aspects, cybersecurity and performance, IP fragmentation must be avoided, and it can only be achieved if you have systems that can automatically adjust MTU to prevent fragmentation or you have to align them manually.

In TCP segments, they use the Maximum Segment Size (MSS), which is calculated by:

IP Packet (1500 bytes) - IP Header size (20 bytes) - TCP Header (20 Bytes) = 1460 bytes.

Let's summarize:

- Each layer has a header and payload; the Ethernet frame of size 1518 bytes has a header of 18 bytes and a payload of 1500 bytes.
- The payload of the Ethernet frame is the IP packet of size 1500 bytes, where 20 bytes for the IP header and 1480 for the payload.
- The payload of the IP header is the TCP segment of size 1480 bytes, where 20 bytes for the TCP header and 1460 for the payload.
- *We use the header as a reference, including the tail (the part added after the payload).*

The use of networking technologies will impact the header size; VLAN, MPLS, IPSec VPN, GRE, and others will consume more bytes and result in less payload size.

Furthermore, each layer has information about the payload type; in the Ethernet frame, it has two bytes for "type," which can indicate if the payload is IPv4, IPv6, ARP, or other protocols. The IP packet header indicates whether it carries ICMP, TCP, UDP, or other protocols within its payload.

Why do we use a byte as a measurement unit?

A letter 'a' in ASCII is represented by the number '97'; ASCII uses a numeric representation of characters, special characters, and numbers. There are many different standards for data representation, such as Unicode, UTF8, and Binary; they all map the data through a table or conversion to one format that can be reversed to obtain the original value. They can all be represented in binary values represented by bits (bit can be 0 or 1). Every 8 bits are called bytes.

In a computer system, information can be physically transmitted over electrical wire in the mean of electrical pulses or wireless by the mean of waves or light pulses in fiber optic. Since we have two possibilities, 0 or 1, it's easy to say 5v equals one and 0v equals zero.

What are the values in OT?

In OT, we have two types of values (set points and tags):

- Digital: 0 or 1 in discrete single bits (serial) or bytes (parallel).
- Analog: constant value output (example, signal or voltage).

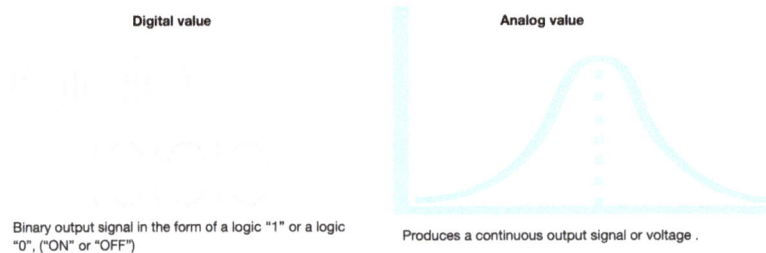

Digital value — Binary output signal in the form of a logic "1" or a logic "0", ("ON" or "OFF")

Analog value — Produces a continuous output signal or voltage.

OT Protocols

Real-time communication protocols are used between different assets of operational technology to read values from sensors (tags), provide values to actuators (set points), obtain information about the health of a device, send control commands (start/stop) and perform maintenance procedures.

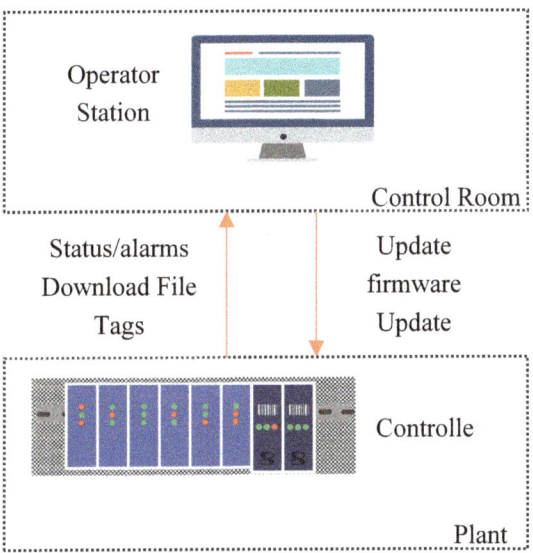

The number of IT protocols used daily is much less than in OT; in OT, there are hundreds of different protocols, and they are widely used in different verticals.

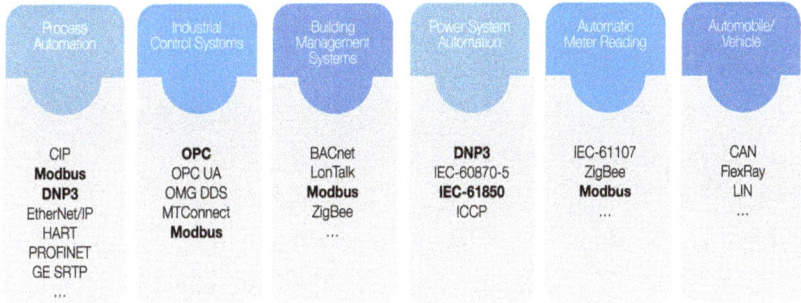

The above is an example of OT protocols; you may notice that some protocols are used across different verticals.

The OT protocols can be TCP/UDP, but not always as non-IP protocols. They will be in Ethernet frame only and no IP or TCP/UDP layers within; this chapter will discuss essential protocols used in different industries.

You need to know that many protocols were developed to be used over serial communications and then developed over Ethernet networks, which may have some limitations; for example, it might have limited payload size because it was initially limited over serial communications and didn't change.

Also, there are different protocols in some verticals in use, depending on which part of the world they may be using other protocols. We have protocols that are similar in functionalities and serve the same purpose.

What to notice as well, OT protocols have initially been developed with no security in mind; most of them lack authentication and encryption.

Another aspect is that most OT protocols are proprietary protocols and not open, making it difficult for cybersecurity controls to provide proper protection since they have no reference. The solution is to collect captures from network traffic running that protocol and analyze them to build a particular parser. Still, there might be some blind spots for the parser because it was only developed based on the collected traffic and not for the protocol as a whole.

The protocols will serve a common purpose but are different in detail; you will notice this while analyzing some protocols in this book.

Tools for network capture analysis are essential; before we begin the study of the protocol, we need to make sure you have all the tools. You also need to download the traffic capture from the download page of www.otsec.io

For viewing the packet captures, I prefer Wireshark, but you can use any tool you are familiar with.

Packet Capture and Network Analysis

The packet capture tools are essential! They listen to frames received on the network card driver level before being processed by the operating system. It's the master tool of network troubleshooting; I jump to capture traffic whenever I troubleshoot communication problems because captures don't lie!

Suppose you have a networking device in between two assets. There is an issue with the communications between them; you had indicators to check if the networking device dropped or altered the traffic. How do you confirm it?

If the networking device doesn't offer enough information for analysis, you will need to start the process of capture-and-compare. And this is not easy to perform, so if you are looking at frames, you have reached a dead-end and decided to look at the traffic to understand the issue.

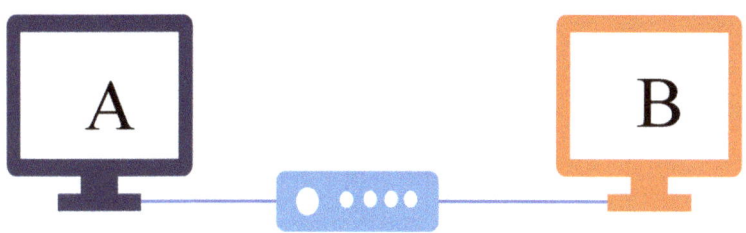

We have four network interfaces involved in these communications 1 at A and 2 at networking device, and 1 at B.

If there is no networking issue, A's payload must be received with no change to B.

Before capturing, you must know that the network devices are usually chatty. The interesting traffic will be mixed with other traffic, so the basic approach is to understand what you are monitoring and apply filters to the capture to only view the concerned traffic.

- *There are two types of filters in packet capture; one applied during capture and one for viewing the capture. If you perform packet capture from CLI 'tcpdump' or Wireshark before starting the capture on a network interface, you can apply a filter that will only collect the matching packets. If you don't apply capture filters, you will have everything on the network interface captured. Then, while viewing the packets, you can apply viewing filters; depending on the use case, you may use both, one, or none.*

Suppose we are troubleshooting Modbus protocol communications, and it's a TCP protocol that works on port 502.

In this case, I will prepare the capture filters as follows:

Source IP = A
Destination IP = B
Destination Port = TCP 502

- This will filter out all other traffic and help focus on the issue.

I need to compare first between the endpoints of the communications so that I prepare the traffic capture on A and B. Then will start the capture on both endpoints and try to reproduce the problem in the shortest time possible from the Modbus application and after the problem occurs, we immediately stop the capture on both endpoints because we need to reduce the amount of captured traffic to make our comparison easier.

Now trace the communications on both captured files and see if they match; if that's the case, most probably nothing is wrong from a networking point of view, and it's an application-level issue.

If you notice a difference, then something was dropped/altered during transmission, so you need to repeat the process of capture and compare on the networking device interfaces; we have an input interface where the traffic is received from A and an output interface where the traffic is forwarded to B.

If you find a difference between the captures of two interfaces of the networking device, then you have spotted a problem! You need to check which frame(s) was changed while processed through the networking device and analyze the cause.

Network troubleshooting is the same in IT and OT, which means the greater the number of nodes on the traffic path between two endpoints (for example, HMI and PLC), the harder it will be to troubleshoot, and this requires more networking knowledge, during this book will discuss important issues that you may face.

During the analysis and depending on the issue, you will need to ask yourself what could this issue be related to?

DNS, DHCP, Routing, Bridging, and others?

For example, you may discover the wrong IP address of what is configured on the application and what is transmitted over the network. That will lead to a list of checks to be performed, such as:

- What is the IP configured on the application?
- What is the IP configured on the OS?
- Is it using DHCP? What is the assigned IP by DHCP?
- Is there a NAT device in between that changed the IP?

- Is it using a Fully Qualified Domain Name (FQDN) instead of an IP? Then DNS is involved!
- Is DNS Server IP configured correctly, either static or through DHCP?
- Is DNS working on the machine, and can it resolve the FQDN to the correct IP?
- Is the DNS server reachable and up?
- Is routing involved, and the endpoints are on different subnets?
- Did you traceroute between the endpoints and confirm the routing is working?
- Is VPN involved and communication with the remote site?
- Is there a possible packet drop due to media errors?
- Are there overlapping subnets in different locations?

The list goes on; the more you're exposed to networking scenarios, the more you will learn and look at exact things faster with time.

How does Modbus send requests over the network?

Modbus is an example, but it's common for all TCP/UDP protocols. As for Modbus, it's a TCP protocol; it begins with a TCP 3-Ways handshake; the first packet is TCP-SYN. Do you know what the processes involved in forming the TCP-SYN are?

There's no payload in the TCP-SYN; it's only the first packet to establish the connection. We need the following information for the TCP-SYN packet:

- Source Mac address
- Destination Mac address
- Source IP Address
- Destination IP Address
- Source Port
- Destination Port

The following are the steps taken place on the sending endpoint to form the TCP-SYN:

1- The application will determine the destination port, if not, it will determine the protocol to use, and the OS will define the destination port number for the protocol.
2- The OS will pick a random source port. The OS will use a range of unused ports to select the source port number, and it's possible to identify the OS from the source port number range.
3- Source IP will be provided from the OS; if the application must use different IP, then there will be a need for other configurations for binding the other IP; it's an exceptional case, in the general OS will select the configured source IP, if OS has multiple IPs then it has a way of picking the correct source IP based on destination IP (routing considered) or based on the order of IP configuration.
4- The application provides the destination IP; if FQDN is in use, OS will check if it has it recorded in a static DNS file or exists in the cache (recently queried). Otherwise, another connection to the DNS server will be established to request the IP address for the FQDN.
5- The OS will provide the source Mac address; if multiple interfaces exist, based on which source IP is configured.
6- OS provides destination Mac address. First, the OS will check if the Mac address for the destination IP is stored in the static ARP records; if not, OS will check if the destination IP is within a directly connected network (in the same broadcast domain) and in this case, it will send an ARP request within the broadcast domain asking who has the "destination IP," usually the destination endpoint, in this case, will answer and provide its Mac address. If the destination IP is not from a directly connected network, then the OS will check if it has any network route to the destination IP; if it's the case, it will use the Mac address of the existing route's gateway as the destination Mac address; if not, OS will use the default gateway Mac address as the destination Mac Address. Of course, if the OS doesn't have the Mac address of the gateway in Static ARP or cache, it will send an ARP request to obtain the Mac address.

Observation:

- You can now imagine how many processes are involved in establishing a simple TCP connection. These steps could be the reason for a problem, and the proper analysis and tracking will help you spot the issue.
- There might be static entries on the OS that will be used as connection parameters; if they are changed by human error or malware, it will cause an issue.
- The Destination IP Address is only used to decide which destination Mac address to use. Communications on the wire are working based on Mac address and not IP; the IP is only used to determine if to connect directly or through the gateway and which destination Mac address to use.
- IT/OT protocols have many details, and communication over an Ethernet network is not straightforward; for an OT cybersecurity professional, you need to realize the level of network communication details, which will also give you an idea about the Cybersecurity flaws of communication protocols.

How can this help an OT cybersecurity professional?

You have probably realized by now that the accepted cybersecurity controls within OT networks are passive, whereas only between levels/zones are active controls such as Firewalls are accepted.

The approach would be to listen to network traffic for analysis and extract features and information about the analysis operation; network analysis is a crucial skill set to obtain, and gathering information by itself is not enough if you are not capable of comparing it to standard or regular traffic to detect anomalies.

We will discuss essential threats to be looking for within the OT network, and following this chapter, we will go through details into a series of commonly used protocols by industry; of course, we can't cover hundreds of protocols within the book, but going through the covered protocols will teach you all the required skills and knowledge to apply the same methodology for other protocols and OT verticals.

Port Scan

It's a method used for active reconnaissance of the network; by actively testing port(s) on an asset to confirm open/close ports, it can also report if there is a firewall or not.

When you run an application that offers a network service, it will listen to the configured port number. Suppose you have a web service running on TCP port 80 on a machine; when the connection begins by TCP-SYN, the device will respond; if there is no service running on the port and it receives a connection, it will send a TCP-Reset to refuse the connection; and if the machine has a firewall blocking TCP-port 80, it will not respond when receiving a connection.

By following this technique, you can report a list of open ports in the network and tell if the firewall is used. Anyhow, ports have a wide range (0–65353), so if you will test all the ports, it means you will have to perform 65K times on a single asset to discover all the ports; it's not realistic, for that reason, network scanner tools by default scan the most commonly used ports, and also, usually the default ports in the tool will be IT ports. The OT protocol ports will be off the radar unless you change the parameter to scan different port numbers.

The network scanning tools work a bit stealthier; the attackers don't want to be detected, so they don't do mass port scanning because the network has cybersecurity controls that can detect their activities. It's also possible that assets have endpoint protection to detect port scans.

The network cybersecurity tools will report the detection of a Syn Scan based on statistics; for example, if they see three or more TCP-SYN to different ports within one minute, there is a port scan.

Over time, network scanning tools have developed beyond the basic port scan; ARP scan, DNS scan, OS fingerprinting, and many others. Also, they have developed techniques to work in stealth mode.

If the scanner, after receiving an answer on the open port, doesn't acknowledge back, it's a flag, so the scanner might complete the three-ways handshake and then close the connection, which makes it harder to detect; they may send some legitimate requests to appear in the logs as regular connection and not only short connection that received no data.

A good example is NMAP, the most known network scanner tool. Nmap ("Network Mapper") is a free and open-source utility for network discovery and security auditing. Many systems and network administrators also find it useful for network inventory, managing service upgrade schedules, and monitoring host or service uptime.

As indicated earlier, we don't tolerate any means of active scanning in the OT network. Active reconnaissance in OT could crash the OT assets and cause a disaster; I must insist on this because it's critical.

Man In The Middle Attack (MITM)

One of the most dangerous attacks on LAN networks. Because it's a simple technique and very effective!

Since we have discussed earlier that communication happens based on Mac address and not IP Address, the MITM attacks work by pretending to own an IP address of the victim.

During the TCP-SYN steps we discussed, ARP (Address Resolution Protocol) is used to obtain the Mac address for the next hop. The Mac address could belong to the destination IP directly or to the gateway to the destination IP, and the ARP protocol is a Layer 2 (Data Link) protocol; it has the source Mac address of the sender and a broadcast destination Mac address "ff:ff:ff:ff:ff:ff" which is received by all Mac addresses within the broadcast domain. The payload of the ARP request includes a question "Who has the Mac address of the IP Address X.X.X.X?" So the host that has the IP Address X.X.X.X will respond to the ARP Request by providing its Mac address.

Now let's take a look at the ARP request/reply on Wireshark.

In Wireshark, there are different windows while viewing packets, and "packet" here and in the term "packet capture" is not limited to IP packets but just used to be called this way most of the time, so don't get confused about that. Packet capture is just a term used by software and file extension (.pcap), but it includes non-IP Ethernet protocols.

In Wireshark, there are three windows: The packet list, the packet details, and raw data.

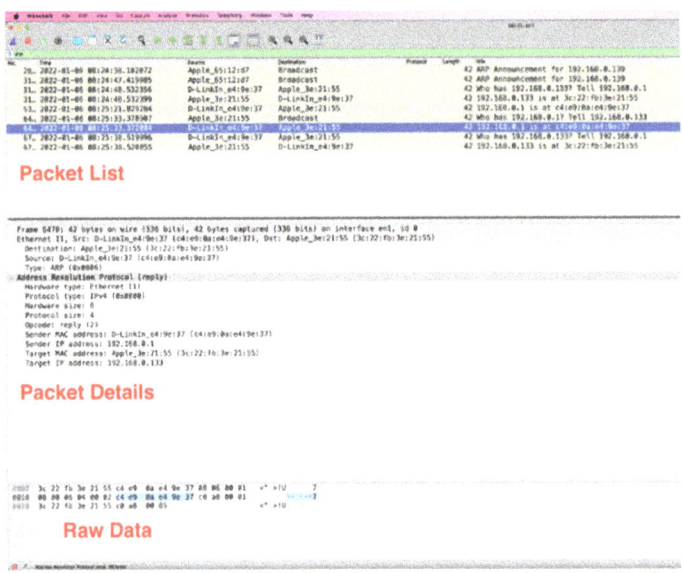

Packet List

Packet Details

Raw Data

We will be looking at the packet details and exploring what details it holds.

The first line is Frame; this is not the actual Data Link Frame; it's Wireshark added header with details on the physical layer.

Ethernet II is the Data Link Frame as per the OSI model, and in the following ARP Request example, you can see it has three details:

Destination Mac address: set to (ff:ff:ff:ff:ff:ff) indicating it's a broadcast.

Source Mac address: the sender's machine Mac address.

Type: The type of the payload is an ARP Protocol in this case.

110

Note that while the ARP is selected in the Packet Details, the Raw Data selects the bytes representing the ARP in the Raw Data window.

You can also read the first unselected bytes, it has the first six bytes for Destination Mac address "ff ff ff ff ff ff," and then six bytes for the Source Mac address, and the last two bytes are "08 06," which is presented in Hexadecimal format as (0x0806), this is the value that represents type ARP protocol.

This gives us the impression that to analyze the protocols, we need to know the order and length of the fields.

Back to the ARP Request:

We have in Packet Details "Address Resolution Protocol (request)," this is a comment by Wireshark about the following ARP section, which includes details as:

Hardware type: Ethernet (1)

The two bytes for the hardware address type will be 1 (0x0001) for the Ethernet network.

Protocol type: IPv4 (0x0800)

The protocol to be mapped to is IPv4, and for IPv4, it is worth always remembering it's (0x0800).

Hardware Size: 6

The length of the Mac address will be six bytes.

Protocol Size: 4

The length of the IP Address will be four bytes.

Op code: request (1)

Type of ARP 1 for request and 2 for the reply.

Then we have Sender Mac address and IP and Target IP, where Target's Mac address is all zeros because it's unknown, and the ARP request is to obtain this value.

The ARP Reply:

```
Frame 910: 42 bytes on wire (336 bits), 42 bytes captured (336 bits) on interface en1, id 0
Ethernet II, Src: Apple_3e:21:55 (3c:22:fb:3e:21:55), Dst: D-LinkIn_e4:9e:37 (c4:e9:0a:e4:9e:37)
Address Resolution Protocol (reply)
    Hardware type: Ethernet (1)
    Protocol type: IPv4 (0x0800)
    Hardware size: 6
    Protocol size: 4
    Opcode: reply (2)
    Sender MAC address: Apple_3e:21:55 (3c:22:fb:3e:21:55)
    Sender IP address: 192.168.0.133
    Target MAC address: D-LinkIn_e4:9e:37 (c4:e9:0a:e4:9e:37)
    Target IP address: 192.168.0.1
```

Now for the reply, it's in the same structure with the following changes:

Op code: reply (2) because it answers the ARP request.

Switch between Sender and Target position because, in case of a reply, the original sender will be the target.

The sender has provided its Mac address along with its IP address.

Observation:

- We need to have references to read the bytes and their meanings.
- The same protocol will be communicated differently depending on the connection stage, and ARP is a simple protocol, but there is more complexity in other protocols.
- The ARP Protocol is clear text to read the contents and analyze them immediately.
- *There was no means of authentication or checking of authoritative answers! Anyone can respond to the ARP request!*

I hope this rings a bell! No authentication, no checking if it's an authorized answer! An attacker in a network can repetitively send ARP replies and force the victim to use its Mac address instead of the actual target's Mac address.

The communication is based on Mac address so that the attacker will receive the packets, and the actual target won't receive them.

What are the possibilities for the attacker after successfully poisoning the ARP table of the victim?

- An attacker can cause **D**enial **o**f **S**ervice (**DoS**), which means the victim will miss all connections coming to it and won't communicate back.
- An attacker can enable packet forwarding capabilities, which means the Victim will be able to communicate back. Still, all requests will go through the attacker, where the attacker has a copy of the traffic and can breach the privacy or alter the communication and modify values.
- The attacker can make the same attack with different victims to force all communications in both directions to go through it. The attacker can have complete control over data and communications.
- We have known incidents where an attacker hijacked communication between control and HMI. The attack modified all control commands from HMI, and the HMI turned as Read-Only with no possibility of having any control.
- Alternatively, I have demonstrated how MITM can hijack the connection and provide fake information to HMI and Control Room when the operation could be in dangerous circumstances. The operators' stations were showing everything was normal.

MITM attack is usually achieved by the ARP poisoning; attackers will try to remove any trails after executing an attack, so they will use ARP protocol to provide the correct Mac address when they are done; this way, they are hoping no one would notice that there was an attack and also operation will appear to be working as usual.

Suppose you have packet capture while the attack occurred; you need to trace when the ARP poisoning started and when it was cleared; it's essential for estimating the damage and analyzing what happened during the attack.

Furthermore, there are other ways for MITM; using Layer 3 Routing poisoning is also possible, and attackers can use the **ICMP** protocol to achieve that.

ICMP Protocol

The ICMP Protocol is famous for the command "ping," but ICMP is a complete protocol suite that has much more capabilities than only "ping" or what is an "echo request."

Internet Control Message Protocol (ICMP), a Layer 3 protocol, it's not a TCP/UDP protocol; it's an IP Protocol.

Mainly used to determine if test connectivity and reachability between IP network devices, it used a technique for sending a message to the destination and calculating the time until the reply to the message is received; it will provide you with information if you have connectivity between two devices and the network performance between them in term of latency.

ICMP has a simple format; it has eight bytes header and a variable payload length.

1 Byte: Type

One Byte: Code

Two Bytes: Checksum

Two Bytes: ID

Two Bytes: Sequence number

Variable length of DATA

Check the following ICMP Echo Request (ping):

```
Internet Protocol Version 4, Src: 192.168.0.133, Dst: 192.168.0.1
Internet Control Message Protocol
    Type: 8 (Echo (ping) request)
    Code: 0
    Checksum: 0x673e [correct]
    [Checksum Status: Good]
    Identifier (BE): 32036 (0x7d24)
    Identifier (LE): 9341 (0x247d)
    Sequence Number (BE): 0 (0x0000)
    Sequence Number (LE): 0 (0x0000)
    [Response frame: 6]
    Timestamp from icmp data: Jan  8, 2022 14:14:49.680605000 +04
    [Timestamp from icmp data (relative): 0.000039000 seconds]
    Data (48 bytes)
        Data: 08090a0b0c0d0e0f101112131415161718191a1b1c1d1e1f202122232425262728292a2b…
        [Length: 48]
```

The type is 8, which indicates an echo request.

In the case of an echo reply, the type will be zero:

```
Internet Protocol Version 4, Src: 192.168.0.1, Dst: 192.168.0.133
Internet Control Message Protocol
  Type: 0 (Echo (ping) reply)
  Code: 0
  Checksum: 0x6f3e [correct]
  [Checksum Status: Good]
  Identifier (BE): 32036 (0x7d24)
  Identifier (LE): 9341 (0x247d)
  Sequence Number (BE): 0 (0x0000)
  Sequence Number (LE): 0 (0x0000)
  [Request frame: 5]
  [Response time: 7.625 ms]
  Timestamp from icmp data: Jan  8, 2022 14:14:49.680605000 +04
  [Timestamp from icmp data (relative): 0.007664000 seconds]
  Data (48 bytes)
    Data: 08090a0b0c0d0e0f101112131415161718191a1b1c1d1e1f202122232425262728292a2b…
    [Length: 48]
```

What are the other uses of the ICMP protocol?

Many IP protocols use the ICMP protocol to communicate messages; it can be used for advertisements, error messages, and queries.

Let's check the possible types/codes:

TYPE	CODE	Description
0	0	Echo Reply
3	0	Network Unreachable
3	1	Host Unreachable
3	2	Protocol Unreachable
3	3	Port Unreachable
3	4	Fragmentation needed but no frag. bit set
3	5	Source routing failed
3	6	Destination network unknown
3	7	Destination host unknown
3	8	Source host isolated (obsolete)
3	9	Destination network administratively prohibited
3	10	Destination host administratively prohibited
3	11	Network unreachable for TOS
3	12	Host unreachable for TOS
3	13	Communication administratively prohibited by filtering
3	14	Host precedence violation
3	15	Precedence cutoff in effect
4	0	Source quench
5	0	Redirect for network

5	1	Redirect for host
5	2	Redirect for TOS and network
5	3	Redirect for TOS and host
8	0	Echo request
9	0	Router advertisement – Normal router advertisement
9	16	Router advertisement – Does not route common traffic
10	0	Route selection
11	0	TTL equals 0 during transit
11	1	TTL equals 0 during reassembly
12	0	IP header bad (catchall error)
12	1	Required options missing
12	2	IP Header bad length
13	0	Timestamp request (obsolete)
14		Timestamp reply (obsolete)
15	0	Information request (obsolete)
16	0	Information reply (obsolete)
17	0	Address mask request
18	0	Address mask reply
20–29		Reserved for robustness experiment
30	0	Traceroute
31	0	Datagram Conversion Error
32	0	Mobile Host Redirect
33	0	IPv6 Where-Are-You
34	0	IPv6 I-Am-Here
35	0	Mobile Registration Request
36	0	Mobile Registration Reply
39	0	SKIP
40	0	Photuris

After looking at this table, you should be worried about allowing ICMP communication on the OT network; unless you have cybersecurity controls that can filter what type of ICMP messages to be allowed, don't allow the whole protocol.

To make it more relevant, when you allow all ICMP messages, it's similar to allowing all TCP port communications. Instead, you must build the ACL for specific source/destination hosts/ports; similarly, you must enable only an ACL for source/destination and the required ICMP message types.

Back to MITM, it can be performed on L3 by using "ICMP Redirect for host or network," where attackers can force communications to go through a different gateway.

ICMP also can be used to perform different types of attacks; we are always worried about simple ones that have a considerable impact. For example, in IT networks today, workstations have 1Gbps network interface cards; if you try to flood the machine using a network flood of 100Mbps rate, it will cause a minor issue, and the workstation might be overloaded in the memory. It may slow down, but it will crash if you do the same load on a PLC!

Ping of Death

You can use the parameter of the ping command to send a large payload; the possibility initially existed for the sake of testing the performance but then was used as an attack; this attack is no more a real threat to today's IT machines because they have enough RAM that can handle this large payload, and only through continuous and from different sources simultaneously can bring the machine down.

What will happen is that this large payload will have to be fragmented, and once reassembled by the victim, it will cause a buffer overflow.

In OT, most assets are sensitive, and such overload will crash them.

ICMP Flood Attack

It's about the size of the payload and the rate; if the attacker is sending hundreds of ICMP to overload the machine, the victim has to process hundreds or thousands of ICMP messages in a second and needs to respond.

ICMP Smurf Attack

Attackers can also spoof their IP addresses, so when the victim replies, it will not go to the attacker's machine; instead, it will also target another victim.

These are simple attacks based on ICMP echo/request where there are many other possibilities by using the ICMP protocol. If there are compromised IoT devices, they can be the source of **Distributed Denial of Service** (**DDoS**), internally and externally, if they have access.

When DDoS is happening, the targeted machine might not crash, but the impact has already happened; the attack consumes the network bandwidth, so other devices probably can't communicate with the target because the bandwidth is consumed. On the other hand, when there's an overload on the target machine, it may not behave normally; some issues will happen to the running services and might also cause indirect damage.

Microsoft Network Services

MS Windows is found in every OT Network, and the network services that MS offers are widely used; due to the popularity of MS OS and applications, it's often the most targeted by cyberattacks. MS will release new cybersecurity patches every second Tuesday of every month. This is indeed a considerable work done by Microsoft, and also there are urgent patches that are released for critical vulnerabilities that can't wait for the following schedule.

What are the services to expect in an OT Network?

You can expect to see all of them, but not all of them will be used, and part of the inventory to build is to include the active services in the network.

MS OS will tolerate active scanning, but if you are going to enumerate all active MS services through active scanning tools such as Nmap, you have to be careful and selective, and scans must not be performed massively; it has to be planned with all teams of the OT Network and Security, and accordingly, you can conduct the scan.

What is usually observed in OT Network are the following:

- File share
- Printing
- Time
- Network Management

We will study the most common IT protocols' services used in OT networks that offer these services.

Server Message Block (SMB)

The SMB protocol is the elephant in the room; I don't recall monitoring any OT network without seeing the SMB protocol in use. They are widely used for sharing files between users in the network.

SMB is used for other purposes, such as sharing printers and serial ports; if you have a device directly connected to a windows machine on a serial port, it can be shared on the network using SMB protocol. Other users can access the device remotely.

SMB allows users to have read/write access to a remote file system; permission is usually set by the resource owner or the network administrator in the active directory. This indicates that SMB has some cybersecurity features embedded.

SMB sessions are usually long-term, established, and users must be identified and authenticated to have access to the shared resource.

SMB is used in Windows systems, whereas the open-source protocol 'Samba' is used in Linux. It was initially developed by IBM in 1984 (SMBv1.0) to share files in Dos (Microsoft Disk OS, a CLI-based OS). Later, in 1996, Microsoft developed SMB1/CIFS (Common Internet File System) for Windows 95, which could support larger file size, Windows RPC service, and NT domain service. Then, followed by the following releases:

- SMBv2.0 was released in 2006 with Windows ME and Windows Server 2008, which could support sharing resources over a WAN.
- SMBv2.1 was released in 2010 with Windows 7 and Windows Server 2008 R2.
- SMBv3.0 was released in 2012 with Windows 8 and Windows Server 2012; it supported encryption and enhanced performance.
- SMBv3.02 was released in 2014 with Windows 8.1 and Windows Server 2012 R2.
- SMBv3.1.1 was released in 2015 with Windows 10 and Windows Server 2016, mainly addressing the cybersecurity issues.

SMB is used for many applications, not only for users to share resources, but many services can use SMB; for example, MS Hyper-V uses SMB to store virtual machines' files. MS SQL Server uses SMB to store user database files on an SMB share.

SMB was made routable by encapsulation in the NetBIOS Protocol. Still, with the introduction of Windows XP and Windows Server 2000, it was implemented over TCP port 445, it works in Client-Server mode, where the client sends a request, and the server responds.

SMB also has many transport implementations over TCP, NetBIOS over TCP, QUIC, or RDMA.

Within the same environment, you can have different versions of Windows and SMB, so how is this handled?

SMB works in three steps:

1) Negotiate the supported SMB version (dialect); the negotiation will either reject a connection due to an unsupported version or agree on a specific version (dialect).
2) After negotiating the protocol, the client will authenticate using the MS SPNEGO protocol, either Kerberos or NTLM.
3) Access resource command indicates the type of operation following successful authentication, which could be as follows:
    - Share access (TREE_CONNECT, TREE_DISCONNECT)
    - File access (CREATE, CLOSE, READ, WRITE, LOCK, IOCTL, QUERY_INFO, SET_INFO, FLUSH, CANCEL)
    - Directory access (QUERY_DIRECTORY, CHANGE_NOTIFY)
    - Volume access (QUERY_INFO, SET_INFO)
    - Simple messaging (ECHO)

Different dialects (versions) have various supported commands; you will find commands for encryption (TRANSFORM_HEADER) and compression (COMPRESSION_TRANSFORM_HEADER) in new versions but not in old ones.

The SMB Protocol is used in many operations. It mounts the remote resource as a local resource, which makes all the usual local operations on the same machine possible on the remote resource, which means it involves many complex operations and possibilities.

When I studied the SMB protocol, I read about 500 pages of documentation from Microsoft; it's a complex protocol, which makes it also vulnerable; many Ransomware attacks were targeting vulnerable versions of SMB; when you indicate the use of an old version of SMB, you must immediately check if it's possible to be updated to the latest version. If that is not possible, then you need to build a cybersecurity policy to restrict and limit access on a mandatory basis considering Zero Trust policies.

NetBIOS

Network **B**asic **I**nput **O**utput **S**ystem (NetBIOS) is not considered a networking protocol but a networking service that enables applications on different machines to communicate in LAN networks. There have been old implementations for token ring networks and IPX/SPX, but with the Microsoft implementation, it's working with IPv4 over TCP/UDP.

In Windows, each machine has a NetBIOS name of 15 characters different from the computer name; the **NetBIOS name** is accessible over **UDP port 137**.

It provides two communication modes:

- **Session mode**: establishes a connection between two computers where large messages can be exchanged with the possibility of error detection and recovery. The client application sends a "call" to another computer (server) over **TCP port 139**; they exchange "send" and "receive" commands until the end of the session. A "hang-up" command is used to terminate the session.
- **Datagram mode**: connectionless, small messages are sent independently, and Applications are responsible for error detection and recovery. The application will listen on **UDP port 138** to receive NetBIOS datagrams; the datagram service sends/receives/broadcasts.

When monitoring OT Network traffic, expect to see NetBIOS traffic; identifying and justifying this type of traffic is essential; if, after review, there is no known application to use NetBIOS, then it's better to be restricted, and of course, if it's known to be used, restrict on a mandatory basis and zero-trust.

Network Time Protocol (NTP)

Don't overlook NTP; it's used to synchronize the time of the machines with the NTP server; if it's compromised, it can cause severe issues to OT applications.

Many operations are running on schedules, and they depend on time. If it is inaccurate, it will break the schedule and corrupt the operation.

Authentication or certificate validation will not work if there is a time difference. All OT communications, audit trails, and logs would mislead and provide wrong indications if the time stamp was incorrect.

NTP synchronization is up to a millisecond level; this is not enough for some types of operations, for example, in power substations; as we will discuss later, there will be the use of Precision Time Protocol (**PTP**), which can provide synchronization up to a nanosecond level.

Another presentation of NTP is the Simple NTP (SNTP), which has no security capabilities at all, where NTP supports authentication using the TLS protocol and synchronization can be protected with encryption and is more reliable as it can sync with several master clocks.

PTP is more advanced than NTP; it has more security capabilities for authentication and encryption and support over complex architecture; it also has redundancy and monitoring capabilities.

Simple Network Management Protocol (SNMP)

It provides management and monitoring for the network devices (switch, router, server, firewall, wireless AP, and controllers).

Network devices that support SNMP have an agent that maintains a database that describes the device's parameters. The maintained information is shared between the agent and the Network Management System (NMS), and the shared data is called the Management Information Base (MIB).

The MIB is a set of parameters and their values; each parameter has a name, Object Identifier (OID), and value.

Ex: name: 'XYZCPU' OID '.1.3.6.1.6.98.1.32.8' value: '89.'

The OID will be in the dotted number format, and the value can be text, count, or number, and is called 'Scalar' representing a single value. Or it can be multiple values and called 'Tabular.'

Every object in MIB is organized hierarchically, which can also be represented by a tree.

It would be best to refer to the network device vendor; usually they have the MIB list published on their website or customer portal because the OID/Names are different from one vendor to another.

The network management system (Manager) uses a set of commands to query MIB:

- GET, to retrieve one or more values
- GET NEXT; get the next MIB in the tree
- GET BULK, retrieve data from a large MIB table
- SET, used by the manager to set values of MIB

SNMP Agent also uses a set of commands:

- TRAPS automatically sends MIB to the manager
- INFORM, similar to TRAPS but has confirmation from the manager on receiving the MIB
- RESPONSE, it the answer to manager GET

SNMP Manager communicates with agents over **UDP port 161**, and when agents are sending unsolicited traps to manager, it uses **UDP port 162**.

There are three versions:

SNMP v1: First version with authentication through a pre-shared key called "community string."

SNMP v2c: Authentication by "community string," added support of 'INFORM' and 'GET BULK.'

SNMP v3: Improved Authentication and Encryption.

One of the common cybersecurity challenges with SNMP when it's using version 1 or 2c is that they don't support encryption; hackers can obtain too much information about the network if they can monitor the traffic.

And also, a common mistake used by many network admins besides not changing the default login credentials is to keep the SNMP community strings on default or commonly used values; it permits attackers not only to obtain information but also to modify the configuration on the network device.

Telnet

Telnet provides access to a command-line interface (CLI) on a remote host. However, due to serious security concerns, it must be disabled and restrict the use of Telnet at any cost.

Telnet is clear text, login credentials can be read by sniffing the network traffic in clear text, and unfortunately, it's still running on many networking devices and IoT devices.

You have to report Telnet as a risk on the network and provide urgent instructions to disable the service, restrict access, and use SSH.

Secure Socket Shell (SSH)

The SSH Protocol provides secure access to a remote host CLI; it offers strong password and public key authentication while all the communications are encrypted.

The SSH service has a list of supported hashing and encryption algorithms; during the connection, the client will negotiate the matching hashing and encryption algorithms with the server and establish the secure shell.

Usually, the SSH service will be running on **TCP port 22**, and it's accepted to be found running on network devices because network admins need to have CLI access to routers, switches, firewalls, and Linux servers. However, there's a 'but' in the sentence.

SSH offers to the tunnel; let's take the following example:

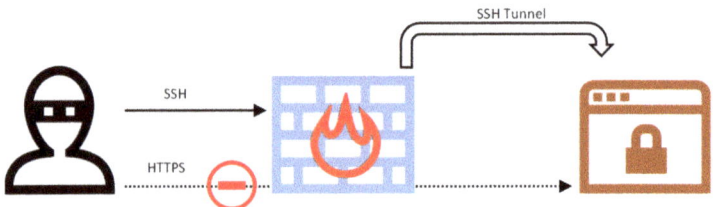

A Firewall is restricting access to web service in the DMZ to a list of authorized hosts/users, if the attacker is trying to access the web service on HTTPS, the firewall will block the connection if the attacker managed to get SSH access to the firewall he can use a feature of SSH to tunnel to any host behind the firewall on any port (HTTPS in this example), and in this case, even if the ACL exist to restrict access, the attacker can still have access.

If the attacker has access to firewall SSH, you may think he can disable the ACL anyway, but what if the SSH is authorized on a Linux server?

Suppose the firewall is allowing SSH to a Linux server but restricting Modbus access to a PLC:

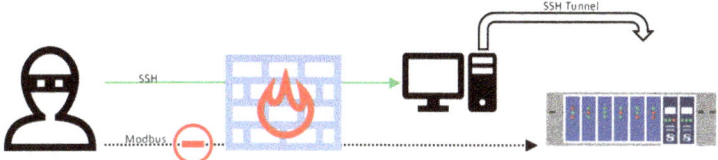

If the attacker managed to gain SSH access to the Linux server, the ACL implemented on the firewall to block any other connection would be useless because the attacker could tunnel to any port on any Linux server host and have Modbus access directly to the PLC.

As a result, you may realize that although SSH is secure transport, it can be a starting point to initiate an attack. This means it has to be secured, and the following are essential points to consider when allowing the use of SSH in the OT network:

- Change the default port of SSH to a random number that is only known to authorized administrators.
- Disable the use of password authentication and only allow public key authentication.
- Disable Root Login.
- Reduce the session idle time to 5 minutes, for example.
- Limit authentication attempts to protect against brute force attacks (dictionary attacks).
- Disable x11 forwarding to prevent access to the web GUI through tunneling.
- ACL must be implemented network-wide to prevent SSH access except for authorized administrators.
- Enable only strong encryption and hashing algorithms.

Secure Sockets Layer (SSL)

SSL is an encryption transport protocol, the protocol is old, and the new updated version of SSL is called **Transport Layer Security** (**TLS**), so don't get confused as SSL and TLS are the same, but TLS is the new updated version.

Since it is a secure transport protocol and the vast majority of network traffic is encrypted by SSL, it must be constantly updated when a security bug is detected. As I am writing this book, the latest version is TLS v1.3, which has dropped the support of weak encryptions and improved performance.

SSL runs on **TCP port 443** when the HTTP Protocol is encrypted by SSL (HTTPS); it also uses port 443. However, when SSL encrypts other clear text protocols, they will be using different ports.

SSL can be used in VPN tunneling (Open VPN), which offers remote access to a network. The user who connects over SSL VPN can access any host on any IP in the private network unless ACL is implemented to restrict the access.

The Open VPN tunnel will assign a virtual IP to the client, and the packet will be formed by using this virtual source IP; then, the whole packet is encrypted and will have another header, so the payload must be reduced to avoid fragmentation in the case of SSL VPN.

Open VPN has suffered from severe vulnerabilities over the years. If it's used to provide remote access to the OT network, then make sure to use the latest updated version and restrict the access to specific resources and services within the agreed schedule; it will be discussed later.

SSL contains two separate protocols:

- Handshake protocols for identifying and authenticating the server because you need to avoid MITM attacks; it will negotiate encryption algorithms and generate a session secure, shared key.
- Record protocols will use the shared key for each session and isolate each session individually.

During the handshake, when a connection is established toward a server, the server will present its server certificate. The client will validate the certificate following the trust chain shown in the certificate:

- Is the certificate valid (start and expiry time and date)?
- Is it issued by a trusted Certificate Authority (CA)?
- Is the certificate name matching the hostname you're accessing?

You may have experienced landing on a warning page when you tried to access a website; the warning will indicate there is a trust issue with the website you're accessing and will state the reason; for example, it may suggest that it's a self-signed certificate or not signed by a trusted certificate authority, and at your own risk, you will accept this warning and proceed to the website.

There is an updated list of trusted Certificate Authorities (CA) in each machine. If you have a local Public Key Infrastructure (PKI) and hence a local CA, it must be added to the list of trusted CAs on each machine to avoid the warning.

You might need to do so when you have a firewall performing SSL Inspection, where the firewall will perform MITM to decrypt the connection and inspect the traffic in clear text. In this case, when the client connects to a server, the firewall will present its certificate claiming the identity of the server and, on the other hand, will establish a secure connection with the server, so you will have the firewall encrypting traffic between client and firewall using the firewall's certificate and encrypting traffic between firewall and server using the server's certificate, if the CA used by the firewall is not trusted in the client machine, it will present a warning.

SSL offers secure transport only; if the traffic encrypted by SSL is malicious traffic, SSL will transport it to the target securely, so SSL doesn't mean it guarantees security. If you have a compromised Engineering Workstation (EWS) by a malware infection, it may try to spread across the network using SSL-encrypted protocols.

There have been some discussions in the OT communities around implementing SSL/TLS into the OT protocols, but I am not a fan of this approach; from where I see it, what is important is the authenticity and integrity of the communications on the low levels and not confidentiality, the use of SSL/TLS for identification is excellent, but we have to consider the following:

- As discussed, SSL will encrypt the data regardless it's good or bad.
- SSL will impact performance; more headers and encrypted data are significant, and processing of encryption/decryption takes longer and consumes resources.
- Cybersecurity controls will not be able to inspect the communication because they are encrypted.
- PKI requires expertise to manage within the OT network.

My conclusion is not to use SSL/TLS for encryption at low levels, but it might be debatable, and I could be missing other use cases worth doing your research about it.

Domain Name System (DNS)

The DNS protocol will translate a domain name into an IP address; in low levels of OT network, it's always preferred to be static IP configuration and avoid the use of FQDN (Fully Qualified Domain Name); however, if for some services that were configured to obtain an IP address through DNS, we have to make sure to restrict access over DNS **UDP port 53** to the authorized DNS server in the DMZ.

DNS can direct the victim to the wrong host; many attacks could cause DNS cache poisoning, where the attacker will respond to the DNS queries with malicious IP addresses.

DNS Protocol format will consist of:

- Headers: Contain information about the number of queries and the query identifier (ID) and other flags.
- Flags: To identify the type of query
- Question: It contains the domain name of the query and record type (A, AAAA, MX)
- Answer: The DNS answer

There are many ways to provide a spoofed answer; MITM can be an easy way to alter the DNS answers; in some cases, it might be about guessing the query ID and spoofing an answer, where you will notice duplicate answers on the network.

DoS attempts can be made by providing extensive answers that exceed the standard size of the fields, and there are many known vulnerabilities in DNS that can lead to the same and cause a buffer overflow.

You need to identify whether or not DNS is used at low levels, justify the use case, and restrict access to only authorized DNS servers.

Summary

In summary, when attacks can be performed by simple mechanisms and tools that cause a significant impact on the OT network, it has to be avoided by any means.

Understanding the communication protocols in the OT network and performing a risk analysis is mandatory; the cybersecurity solution you are going to propose has to take into account all these aspects.

It's not an easy mission, but many cybersecurity controls have the capabilities to handle these threats; you need to have an inventory of all possible threats based on the OT network communication protocols, including both IT/OT protocols, and then make sure you have the proper mitigation plan.

Part 3: Common Industries

Chapter 9: Electrical Power Substations and IEC61850

Before we discuss this subject, let's have a quick brief around essential terms, only to simplify them and make them relevant during the discussion.

We have two power sources, AC and DC, so what are they?

DC: Direct Current, it's a one-direction flow of electric current constantly.

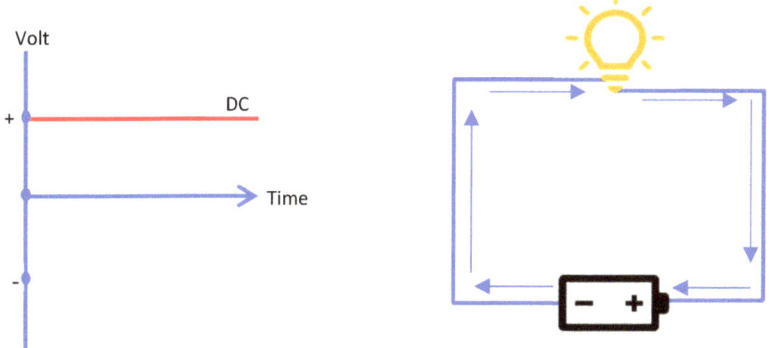

Any device running on a battery is an example of DC, Thomas Edison was the advocate of DC and tried to push it as the primary power source, but the problem was that with DC. It isn't easy to be used in transmission. High voltage DC, if transmitted over long distances, will have high resistance from the wires and there will be power loss; the ordinary wires we see today will melt if used to transmit High-voltage DC. But DC is still used. Usually, everyday home appliances will be working on DC.

AC: Alternating Current is a flow of electrical current back and forth at regular intervals called cycles; AC is used for transmission since the voltage can be wide-ranging by using transformers. Nikola Tesla was behind pushing for the benefits of AC for electric power transmission; I invite you to check the history of Nikola Tesla and the debate he had with Thomas Edison.

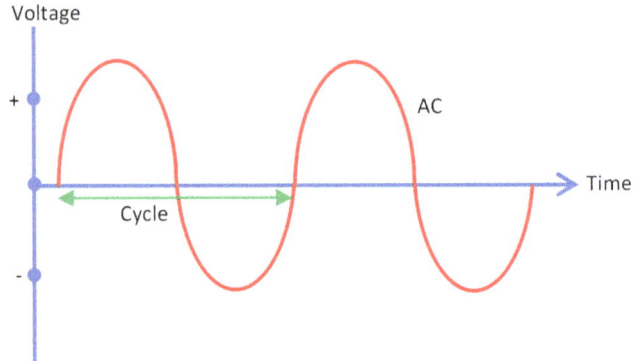

The number of cycles per second is called frequency and measured in (Hz); you will find that appliances support 50Hz or 60Hz depending on which country you are in.

So if you have a light working on AC power at 50 Hz, it means it goes on and off 50 times in a second, but the human eye won't be able to notice the difference.

The electricity will be transmitted over AC to your home, and the appliances will use an electric device called a "rectifier" that reverses the direction of (AC) current at the proper frequency to make it steady (DC).

What are electric current and voltage?

Current is the flow of electric charge, negatively charged electrons from one place to another over an electric conductor (ex: copper wire). It's measured by Amperes (Amps); we see on various electronic devices the supported Amps because these devices can handle a certain amount of electrons to pass through them and if it exceeds that, it will burn.

Voltage (V) measures the force that moves the current; if you need more current, it means you need more voltage.

Let's take a look at the following example that you may be familiar with to be found on power adaptors:

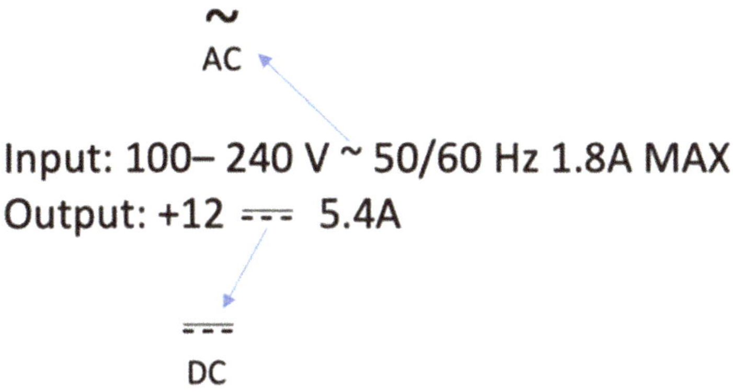

AC ~

Input: 100– 240 V ~ 50/60 Hz 1.8A MAX
Output: +12 ⎓ 5.4A

DC ⎓

It indicates that the power adaptor takes a source of AC power of voltage range (100–240 V), supports 50Hz and 60Hz frequencies with a maximum of 1.8A, and provides 12 V DC output at 5.4 Amps.

Transformers

Electrical measuring devices:
Current Transformer (CT): Lowers (steps down) the current.
Potential (or Voltage) Transformer (PT): Reduces high voltage to low voltage

Disconnectors

Used to ensure that an electrical circuit is completely de-energized for service or maintenance.

Circuit Breaker

Electrical safety device designed to protect an electrical circuit from damage caused by an overcurrent or short circuit.

Merging Unit

Measures signals from current and voltage transformers and merges them into digitalized output that can be provided to protective devices (IED).

After having these few terms from our way, let's discuss how the power is generated, transmitted, and distributed.

Power Generation

It is the process of generating electric power from various energy sources using chemical or mechanical processes.

The most economical power generation today is using mechanical methods based on Faraday's law of induction. if you rotate a magnet between two coils, for every rotation, the magnet will move the electrons forward and backward per the magnet poles rotations, which generate an AC current.

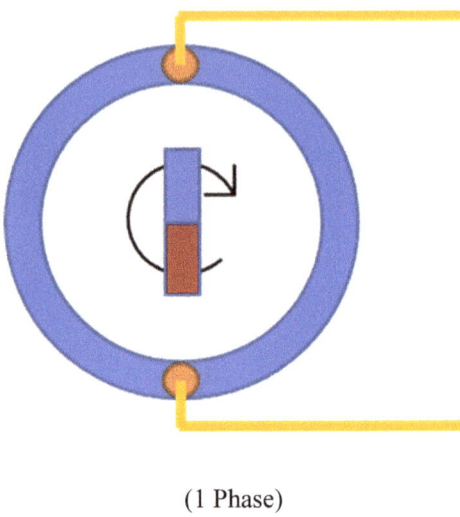

(1 Phase)

When we have two coils, it's called 1 Phase; the most used model is the 3 Phase, by adding another two pairs of coils to use the single rotation to generate more AC current efficiently.

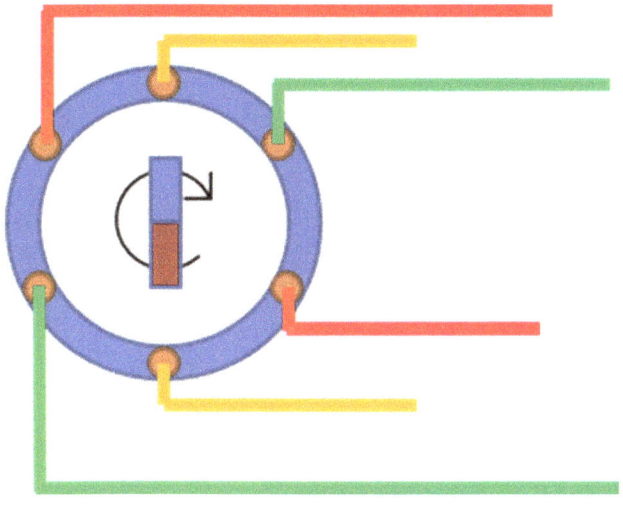

(3 Phase)

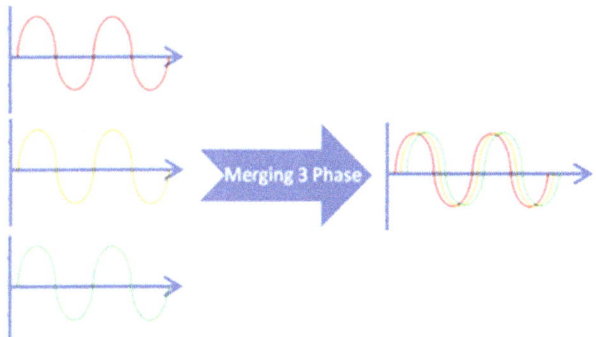

How to get the mechanical rotation?

Today's industries use steam power to create rotational movement. This can be generated by heating the water to make the steam and then cooling it down, which requires fuel for this operation, and usually, it can be generated by coal, natural gas oil, or nuclear power.

It's also possible to use the natural flow of water or use a water dam, and there are wind power and many other clean natural sources.

Power Transmission

The process of moving power from generation plants to electrical substations. The interconnected lines that facilitate this movement are known as the transmission network, an efficient long-distance network that carries high voltage.

Power Distribution

The final stage of electric power delivery carries the power from the transmission network to individual customers.

Transformers reduce the high voltage from the transmission network to medium voltage (2–35 kV). Then, distribution transformers near customers' premises lower the voltage to consumer range voltage.

The transition from transmission to distribution happens in power substations, which have components such as transformers, disconnectors, and circuit breakers.

The network of interconnected lines and components from generation through the transmission to distribution is called power grid.

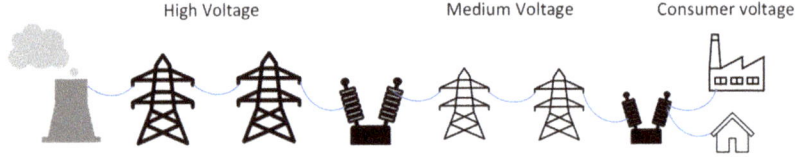

Power Grid

The power grid is a vast operation, and you will find different teams responsible for generation, transmission, and distribution.

Within the generation plants, it's an operation that produces electricity as its final product, wherein transmission, it's focused on the delivery of high voltage. In distribution substations, there are different aspects of the operation. It elevates and reduces the current and voltage; it distributes to consumer substations, an advanced process that must be monitored and controlled.

Electrical power systems are vital for our lives, and modern life requires an uninterrupted power supply. The power demand is high, and more substations are added to the grid to have more expansions.

The management of these time-critical operations is achieved through the Substation Automation Systems (SAS), and the standard for communication is IEC 61850.

What is IEC-61850?

IEC-61850 is an international standard defining the communication protocol for electrical substations for intelligent electronic devices (IED).

IEC-61850 standard has several parts. Each will cover one aspect of the standard; they are listed here, and you just need to take a quick look:

- IEC TR 61850-1:2013 – Introduction and overview
- IEC TS 61850-2:2003 – Glossary
- IEC 61850-3:2013 – General requirements
- IEC 61850-4:2011 – System and project management
- IEC 61850-5:2013 – Communication requirements for functions and device models
- IEC 61850-6:2009 – Configuration language for communication in IEDs
- IEC 61850-7 – Basic communication structure
- IEC 61850-8-1:2011 – Specific communication service mapping
- IEC 61850-9-2:2011 – Specific communication service mapping
- IEC/IEEE 61850-9-3:2016 – Precision time protocol
- IEC 61850-10:2012 – Conformance testing
- IEC TS 61850-80 – Guideline for exchanging information
- IEC TR 61850-90 – Use of IEC 61850 for the communication between substations
- IEC TR 61850-90-2:2016 – Using IEC 61850 for communication
- IEC TR 61850-90-8:2016 – Object model for E-mobility
- IEC TR 61850-90-12:2015 – Wide area network engineering guidelines

Some of these parts also have different subparts with more specifications, and the subject can be described in thousands of pages due to the too many details involved. Still, we will extract the necessary information for your attention and explain what matters to you as an OT cybersecurity professional.

The purpose of IEC-61850

- Address the need for a more structured approach to the design of substation automation systems.
- Separate the data model from the method of communication; utilize new technologies (Ethernet, TCP/IP).
- Enable vendor independence by the possibility of integrating several IEDs from different vendors.
- Simplify system configuration that can be exported and imported in an XML similar format.
- Enable sharing of measurement among devices.

The IEC-61850 Data Modeling

Data will be presented in a hierarchal structure, and all that is required is to map this structure to any other device's structure to access the specific parameters.

Suppose you have a code that defines the following:

Length: X

Width: Y

Color: Z

And you need to integrate with another code that requires reading these values, but they are defined differently:

AA is for length, but the value can't be a decimal and always must be an integer.

AB: is for the width and also must be an integer value.

AC: Color, but it must be all lower case.

When you do the integration between two codes, you need to do mapping and proper conversion, so the other code needs to read information from your code in the following way:

AA = Integer (X)

AB = Integer (Y)

AC = Lower Case (Z)

Based on this type of mapping and conversion, even if you have different systems and presentations, you can find a way to make the two systems communicate.

We have many OEM vendors that manufacture IEDs. Based on features and functionalities, each IED has a set of objects and attributes; instead of making two IEDs from different vendors communicate directly, IEC 61850 defines a data model that can be understood by all IEDs and is also exchangeable.

The Standard doesn't deal with physical devices. Instead, it defines logical devices on the physical device; each logical device has logical nodes.

Logical nodes have a set of logical objects and attributes; in this way, you only need to map the logical objects and attributes between logical nodes on different logical devices.

Even better, the OEM vendors provide template files that can be imported/exported between IEDs with all the data model mapping required, so it's straightforward to do the integration.

For example, when exchanging data or sending commands, it has to be mapped to the right path:

- Logical device
- Logical node
- The function
- Data object
- Data attribute

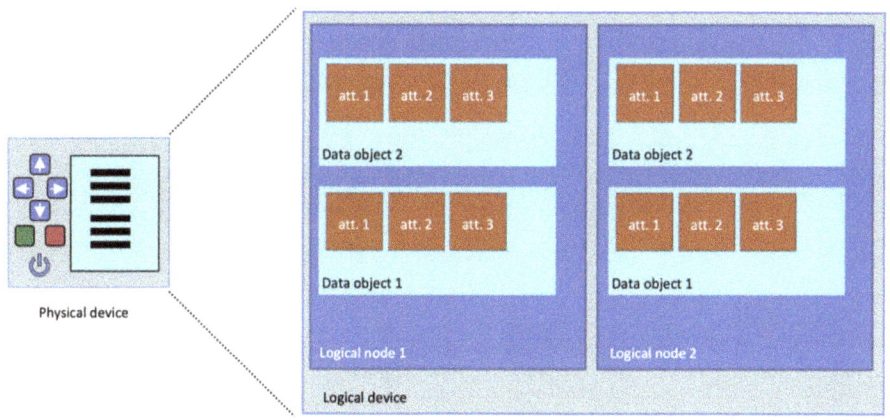

In this way, if you want to read an attribute's value, you have to define a complete path:

Logical Device/Logical Node $ Data Object $ Data Attribute

Example: IED01/XCBR1PosstVal

Logical device name: IED01

Logical node: XCBR1

- *A logical node is linked to a function; logical nodes have defined names and are distinguished by numbers; for example, XCBR is a circuit breaker, and you will notice that the first letter always indicates the function, such as: 'C' for control, 'M' for measurement, and 'P' for protection, and 'X' for switch gear.*

Data Object: Pos, Pos is for the position of the circuit breaker (On/Off)

Data attribute stVal, "st" for the status, and "Val" for the status value.

In this hierarchical way, you can access the exact attribute and function, and by using vendors' template files, you can make IEDs communicate with each other for exchanging data and control.

IEC 61850 Network Architecture

In IEC 61850, we use different naming of the low levels, Level 0, which has devices such as Merging Unit (MU), circuit breakers, and transformers, called Process level. The difference here is that MU interface with other devices and communicate over Ethernet with IEDs; the IEC 61850 protocol in use at this level is Sample Measured Value (SMV) or also might be called (SV), which is a Data Link protocol and works in Publisher/Subscriber method.

In level 1, it's called bay level, which has IEDs connected on redundant links that communicate over the IEC 61850 GOOSE protocol; it's also the Data Link protocol and works in the publisher/subscriber method.

Level 2 is called the station level, which has supervisory machines such as a SCADA server, an HMI, and an EWS. Usually, EWS might be called an IED configurator. The communication between the IED and SCADA server over the IEC61850 Manufacturing Message Specifications protocol (MMS) it's a TCP protocol that works in client/server mode.

Level 2 could be physically located in a remote control room.

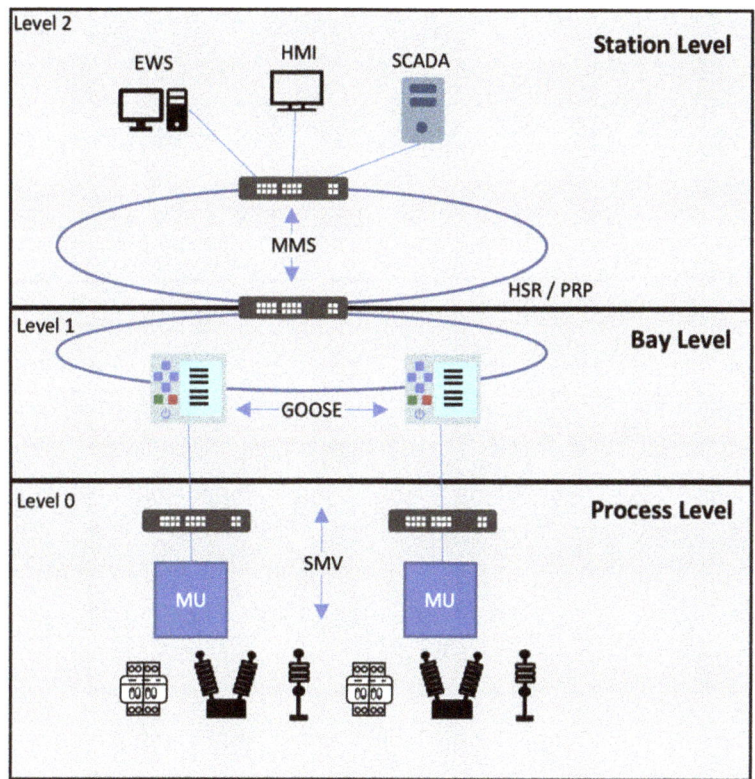

IEC 61850 Protocols

Now we will analyze the three protocols:

- MMS
- GOOSE
- SMV

Manufacturing Message Specifications Protocol (MMS)

Client/server protocol, used for communications between IEDs and SCADA (MTU and RTU). It facilitates access to data in IEDs for read/write and also exchanging of files.

It's not only used in substations and smart-grid, as originally was developed by General Motors for manufacturing, but also used in wind power plants for monitoring and control.

MMS has many methods; the following some examples:

- *Get* to read values.
- *Set* to write values.
- *Query Attributes* to obtain the attributes of an object.
- *Create*, create an instance of objects inherited from a defined class.
- *Rename*, rename and objects.
- *Delete*, and delete an object.

MMS was developed from an old proprietary protocol, and with time it became public and could be used by multi-vendor. MMS is an ISO protocol family and not an Internet protocol family (TCP/IP) – *the specifications of ISO are not available publicly free of charge*; you will find extra headers on top of the TCP header; check the following application layers of MMS:

```
TPKT, Version: 3, Length: 211
ISO 8073/X.224 COTP Connection-Oriented Transport Protocol
ISO 8327-1 OSI Session Protocol
ISO 8823 OSI Presentation Protocol
ISO 8650-1 OSI Association Control Service
MMS
```

The ISO protocol family has a Transport packet (TPKT) on top of the TCP Header. Then COTP or CLTP, session protocol, presentation protocol, and application.

- *COTP: Connection-oriented transport protocol, similar to TCP.*
- *CLTPL Connectionless Transport Protocol is similar to UDP.*

If you are not familiar with the ISO protocol family and try to study an MMS, you will get confused; what you need to know is that it's another standard format that is used to provide transport to the applications; I will not discuss much about the details of the headers. I will only look at the MMS application.

When you trace a connection over MMS, as in the case of TCP, there will be a TCP 3-way handshake; following that, the initiator will send a connection request, and the responder will answer by connection confirmed. Until this point, it's handled by the ISO headers and not the MMS application protocol.

Connect Request:

```
TPKT, Version: 3, Length: 22
ISO 8073/X.224 COTP Connection-Oriented Transport Protocol
    Length: 17
    PDU Type: CR Connect Request (0x0e)
    Destination reference: 0x0000
    Source reference: 0xb001
    0000 .... = Class: 0
    .... ..0. = Extended formats: False
    .... ...0 = No explicit flow control: False
0000  03 00 00 16 11 e0 00 00  b0 01 00 c0 01 0a c1 02
0010  00 01 c2 02 00 02
```

Connect Confirm:

```
TPKT, Version: 3, Length: 14
ISO 8073/X.224 COTP Connection-Oriented Transport Protocol
    Length: 9
    PDU Type: CC Connect Confirm (0x0d)
    Destination reference: 0xb001
    Source reference: 0x1802
    0000 .... = Class: 0
    .... ..0. = Extended formats: False
    .... ...0 = No explicit flow control: False
0000  03 00 00 0e 09 d0 b0 01  18 02 00 c0 01 0a
```

As the TCP communications will set the ACK flag in every communication except in the establishment phase of the connection, it's normal to see TCP ACK sent after any exchange of TCP data.

For MMS, following any data exchange in any direction is followed by:

- TCP ACK
- TCP PSH, ACK

Why set PSH flag?

The push flag is used to avoid TCP delays. The delays in TCP are useful and help in improving the efficiency of communications and guarantee of receiving the data, but in critical applications, the TCP push flag is used to inform the client or server to send the data it has without waiting for the buffer to be filled.

It's not always typical to see a PSH flag, as it might indicate malicious activity, but in MMS, it's normal to see it following every packet exchange.

TCP ACK:

```
Flags: 0x010 (ACK)
    000. .... .... = Reserved: Not set
    ...0 .... .... = Nonce: Not set
    .... 0... .... = Congestion Window Reduced (CWR): Not set
    .... .0.. .... = ECN-Echo: Not set
    .... ..0. .... = Urgent: Not set
    .... ...1 .... = Acknowledgment: Set
    .... .... 0... = Push: Not set
    .... .... .0.. = Reset: Not set
    .... .... ..0. = Syn: Not set
    .... .... ...0 = Fin: Not set
    [TCP Flags: ·······A····]
```

TCP PSH, ACK:

```
Flags: 0x018 (PSH, ACK)
    000. .... .... = Reserved: Not set
    ...0 .... .... = Nonce: Not set
    .... 0... .... = Congestion Window Reduced (CWR): Not set
    .... .0.. .... = ECN-Echo: Not set
    .... ..0. .... = Urgent: Not set
    .... ...1 .... = Acknowledgment: Set
    .... .... 1... = Push: Set
    .... .... .0.. = Reset: Not set
    .... .... ..0. = Syn: Not set
    .... .... ...0 = Fin: Not set
    [TCP Flags: ·······AP···]
```

Following that, the first MMS application PDU is in the form of an "initiating request" that will take place:

```
MMS
  initiate-RequestPDU
      localDetailCalling: 32000
      proposedMaxServOutstandingCalling: 20
      proposedMaxServOutstandingCalled: 20
      proposedDataStructureNestingLevel: 4
      mmsInitRequestDetail
        proposedVersionNumber: 1
        Padding: 5
        proposedParameterCBB: fb00
        Padding: 3
        servicesSupportedCalling: 6e1d000000000064000198
0060  02 06 00 61 60 30 5e 02  01 01 a0 59 60 57 80 02   ···a`0^·  ··Y`W·
0070  07 80 a1 07 06 05 28 ca  22 01 01 a2 04 06 02 29   ······(·  "······)
0080  02 a3 03 02 01 02 a6 04  06 02 29 01 a7 03 02 01   ········  ··)·····
0090  01 be 32 28 30 06 02 51  01 02 01 03 a0 27 a8 25   ··2(0··Q  ·····'·%
00a0  80 02 7d 00 81 01 14 82  01 14 83 01 04 a4 16 80   ··}·····  ········
00b0  01 01 81 03 05 fb 00 82  0c 03 6e 1d 00 00 00 00   ········  ··n·····
00c0  00 64 00 01 98                                     ·d···
```

Following are some examples of MMS PDU:

- Initiate-RequestPDU
- Initiate-ResponsePDU
- Initiate-ErrorPDU
- Confirmed-RequestPDU
- Confirmed-ResponsePDU
- Confirmed-ErrorPDU
- Conclude-RequestPDU
- Conclude-ResponsePDU
- Conclude-ErrorPDU
- Cancel-RequestPDU
- Cancel-ResponsePDU
- Cancel-ErrorPDU

The Initiate-Request will have a proposal of what is supported in terms of version, parameters, and services (methods/functions); it contains lists with values of either true or false.

Each object is attached to a list of possible methods, and they are called services.

Proposed Parameters:

```
v proposedParameterCBB: fb00
     1... .... = str1: True
     .1.. .... = str2: True
     ..1. .... = vnam: True
     ...1 .... = valt: True
     .... 1... = vadr: True
     .... .0.. = vsca: False
     .... ..1. = tpy: True
     .... ...1 = vlis: True
     0... .... = real: False
     .0.. .... = spare_bit9: False
     ..0. .... = cei: False
```

Supported Service Calling:

```
servicesSupportedCalling: 6e1d0000000000064000198
 0... .... = status: False
 .1.. .... = getNameList: True
 ..1. .... = identify: True
 ...0 .... = rename: False
 .... 1... = read: True
 .... .1.. = write: True
 .... ..1. = getVariableAccessAttributes: True
 .... ...0 = defineNamedVariable: False
 0... .... = defineScatteredAccess: False
 .0.. .... = getScatteredAccessAttributes: False
 ..0. .... = deleteVariableAccess: False
 ...1 .... = defineNamedVariableList: True
 .... 1... = getNamedVariableListAttributes: True
 .... .1.. = deleteNamedVariableList: True
 .... ..0. = defineNamedType: False
 .... ...1 = getNamedTypeAttributes: True
 0... .... = deleteNamedType: False
 .0.. .... = input: False
 ..0. .... = output: False
 ...0 .... = takeControl: False
 .... 0... = relinquishControl: False
 .... .0.. = defineSemaphore: False
```

The responder will send initiate-response back, stating supported parameters and services.

Following to initiate-request/initiate-response, the client will send to the device (as a server) a request, for example:

Confirmed-RequestPDU

- This request will provide an identifier called "invokeID," represented in numeric value; it will be used to distinguish and map responses to appropriate requests.
- It will also specify the Confirmed Service Request type, for example, 'getNamelist.'
- This request also needs to provide the object class, for example, 'domain.'
- Then it needs to specify the scope, for example, 'vmdSpecific.'
    o *VMD: Virtual Manufacturing Device (Logical Device)*

Confirmed Request:

```
 MMS
   confirmed-RequestPDU
     invokeID: 1
     confirmedServiceRequest: getNameList (1)
       getNameList
         extendedObjectClass: objectClass (0)
           objectClass: domain (9)
         objectScope: vmdSpecific (0)
           vmdSpecific
```

Confirmed Response:

```
 MMS
   confirmed-ResponsePDU
     invokeID: 1
     confirmedServiceResponse: getNameList (1)
       getNameList
         listOfIdentifier: 5 items
           Identifier:
           Identifier:
           Identifier:
           Identifier:
           Identifier:
         moreFollows: False
```

Note that the InvokeID is: 1

In this example, the response for getNameList provides a list of five VMDs identifiers (covered in black); these identifiers can be used in further requests as references for objects and attributes.

One of these identifiers will request a list of all objects and methods.

Confirmed Request:

```
 MMS
   confirmed-RequestPDU
     invokeID: 2
     confirmedServiceRequest: getNameList (1)
       getNameList
         extendedObjectClass: objectClass (0)
           objectClass: nammedVariable (0)
         objectScope: domainSpecific (1)
           domainSpecific:
```

Service requested: getNameList

The scope specifies a unique VMD name from the previous command, "covered in black."

Confirmed Response:

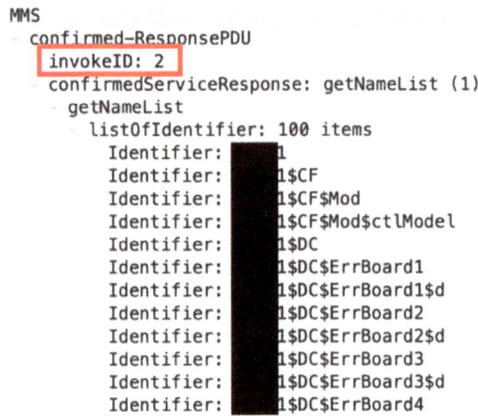

Matching invokeID: 2

Providing the list of 100 items in the hierarchical naming format that was discussed earlier: Logical Device $ Logical Node $ Data Object $ Data Attribute

The following communications can depend on this information to perform several operations on this list of attributes and methods.

Once the confirmed request/response is made, there can be a confirmed-conclude to finalize all remaining requests and tear down the connection.

Let's highlight important notes about MMS:

- MMS is an ISO family protocol.
- VMD is not an actual device; it's the logical device; it hides the real device and provides an abstract of data; multiple VMDs can be found on one real device.
- The role of the client/server is exchangeable depending on the request/response.
- TCP/IP will specify the target device by IP and port; then, the logical device VMD defines access following the IEC 61850 data modeling.
- MMS Connection will begin by "initiate" and terminate by either "conclude" (which will answer all remaining requests) or by "abort" (which will delete all pending requests).

- The objects are defined within scope; the scope can be:
    - Variable name
    - Domain name
    - Connection
- Each object has attributes and is associated with a list of services (ex: read/write/delete/rename).
- A confirmed response or confirmed error will answer each confirmed request.
- InvokeID links confirmed the request.
- A confirmed request can be canceled.
- Another type of request is an unconfirmed request, which doesn't receive a response and can't be canceled. It provides unsolicited status, events, and reports.
- MMS protocol has a broad coverage of services, objects, and attributes, and the subject is extensible. We only covered some basics, but they should make you comfortable analyzing MMS communications.

For further analysis, please refer to the website on the reference page.

MMS Security

As you may have noticed, the communications are in clear text; in the available captures online, I couldn't find any means of authentication or encryption, but as per "IEC 62351-4:2018," it specifies security requirements both at the transport layer and at the application layer. So it's important to check the possibility of implementing this level of security if applicable, and this must be confirmed with the OEM vendors.

Another issue is that the Request/Response method relies on invokeID, making it easy to be replayed or spoofed; all types of MITM attacks can directly impact and take complete control over the communications.

An attacker can obtain complete information about the VMD's object attributes and services available by passively monitoring the traffic and can have unauthorized access and cause damage.

It's essential to analyze the MMS traffic you are supposed to secure. If you found no security implementations exist, you need to check the possibility of upgrading through the OEM vendor to support IEC 62351-4.

If it's not possible, and in anyhow, this will also be addressed in the cybersecurity controls chapters.

GOOSE Protocol

Generic Object Oriented Substation Event (GOOSE).

Within substations, we have different transformers and circuit breakers taped to other voltage lines; for example, if we have one line of 100kV connected to several 10kV lines, if a shock is detected on the 10kV line, it won't be visible at the 100kV line's protection relays. For that reason, it was common to use ground switch. If a shock is detected at the 10kV line, the protection relay will send a trip signal to the ground switch, and then the 100kV protection will see that fault and react as a fault on its line is somehow not efficient and has many drawbacks.

When using a communication-capable device, such as an IED at each relay, it can inform the other IEDs of the fault without the need to cause fault across different lines. There are many applications and use cases that several parts of the substations must be aware of by using the IED's communication capabilities.

IED will facilitate the updates about the status between several controls for many applications and use cases, and the protocol in use between IEDs at the Bay level is GOOSE.

In the above example, an IED will send a Transfer Trip GOOSE message to all other connected stations from the faulty station.

What are the characteristics of GOOSE?

- GOOSE provides secure and reliable communications between substations. It uses the Publisher/Subscriber model, where the publisher can send to multiple subscribers fast and reliable.
- GOOSE sends to unicast or multicast, depending on subscribers.
- It can't utilize confirmed messages as in MMS because processing confirmation of multi subscribers will be time-consuming. In contrast, the time required at the bay level is less than 5ms, unlike at the station level of the MMS case, which can tolerate up to 100ms.
- GOOSE might use VLAN, which helps in prioritizing the communication.
- Instead of message confirmation and to achieve fast and reliable communication, GOOSE will assume the messages were not received by all subscribers and will resend them at a fixed rate.

- If not receiving the data from the publisher as per the expected fixed rate, it can tell that there's a connection loss.
- GOOSE is event-driven; messages will be sent on every new event occurrence.
- Each message contains the time interval for the following message.
- GOOSE is a layer 2 "Data Link" protocol.
- GOOSE message contains:
  o Identification
  o Datasets (object and list of attributes)
  o Time of the subsequent transmission.

GOOSE PDU

Let's take a look at the sample from the Wireshark website:

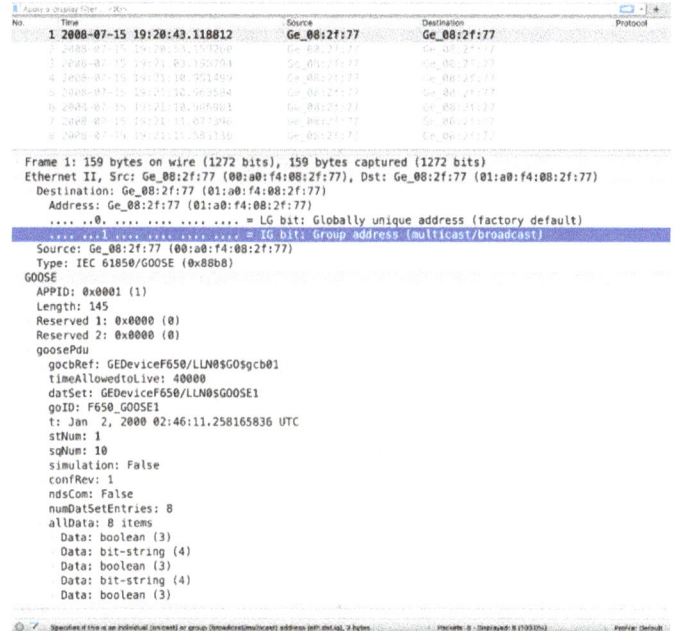

Not that the group address bit is set (multicast).
What is in the GOOSE PDU?

gocbRef	Is the unique identifier reference of the control block
timeAllowedtoLive	Time to wait for the next transmission
dataSet	The objects and attributes list in IEC61850 data model format
goID	IED identifier
t	Timestamp
stNum	Counter of events
sqNum	Sequence number incremented with every resending of a message
Simulation (Test)	Test bit to indicate if it's a simulation message
confRev	Configuration revision number, incremented on every configuration change
ndsCom	Commissioning of configuration when control block needs to be reconfigured
numDataSetEntries	Number of entries in the dataset
allData	List of items and values

GOOSE Security

Considering the time requirements of GOOSE <5ms, it makes it very difficult to add encryption, as encryption has overhead processing and delays. It can only be achieved by powerful processors, which means you may need to replace the hardware to perform encryption without impacting the performance requirements.

There are several working proposals by OEM vendors to provide encryption; to avoid the overhead delays, the essential parts of the message can be encrypted, such as counters and time stamps, and the use of a reasonable cipher key length can still meet the time requirement.

Without authentication and encryption, GOOSE is subjected to all types of L2 attacks, including MITM, spoofing, and replay.

An attacker can send periodic spoofed messages to subscribers to present incorrect data, which could have a severe impact.

A control from switch level and VLANs to isolate the IED devices is mandatory, besides other cybersecurity controls that we will discuss later.

Sample Measured Value (SMV)

SMV is similar to GOOSE; it's an L2 data link protocol and works as a Publisher/Subscriber model. It is used for communications between MU and IED.

Where GOOSE is event-driven, SMV is stream-based. A publisher will send periodic messages at exact, fixed intervals. SMV offers various advantages, such as facilitating the interoperability of data and reducing the complexity of the system connectivity.

SMV time requirements are less than MMS and GOOSE; MMS delays must be < 3 ms.

SMV messages are classified into topics; subscribers will receive all but only process the subscribed topic.

The rate is defined depending on the vendor and use cases and calculated based on samples per second (80, 256, or 30,120 samples/second) and the measured signal frequency (50 or 60 Hz).

Example: Frequency is 60 and samples per second is 30, the rate will be:
$1/60/30 = 65 \ \mu s$

Now let's take a look at the SMV PDU (from github.com/megadelha/):

```
Frame 10153: 120 bytes on wire (960 bits), 120 bytes captured (960 bits)
Ethernet II, Src: ca:fe:c0:ff:ee:69 (ca:fe:c0:ff:ee:69), Dst: Iec-Tc57_04:00:02 (01:0c:cd:04:00:02)
▸ 802.1Q Virtual LAN, PRI: 4, DEI: 0, ID: 1
IEC61850 Sampled Values
  APPID: 0x4001
  Length: 102
  Reserved 1: 0x0000 (0)
  Reserved 2: 0x0000 (0)
  savPdu
    noASDU: 1
    seqASDU: 1 item
      ASDU
        svID: 4001
        smpCnt: 832
        confRev: 1
        smpSynch: global (2)
        seqData: 0000fd5e0000000000031cf000000000fffbe8e60000000000000033400002000003ee39e...
```

Note the 802.1q VLAN tag, and it's been configured for priority class 4.

Each SMV PDU can carry up to 8 3Phase current and voltage measurements, and this is defined as the number of ASDU (Application Specific Data Unit). In the capture, it shows as:

noASDU: 1

Subscribers will receive all ASDUs but will only read the concerned ones.

The list of ASDUs will be numbered in sequence (seqASDU).

The example above shows a single ASDU, but you can find a list of ASDUs in the same format.

Each ASDU has an identifier (svID); it's a user-defined string; subscribers will subscribe to svID.

There is an incremental counter (smpCnt).

Similar to GOOSE, SMV has versioning of the configuration; each configuration change increments the value of (ConfRev) by 1.

smpSync indicates the time synchronization and has these values:

- 1, no sync
- 2, local sync
- 3, global sync

Seq Data is the actual measurements of currents and voltages in sequential format; the subscribers will be configured to parse these data and scale for values, as scaling will differ between voltage and current values.

SMV Security

SMV is very similar to GOOSE, and both are restricted in terms of delays. However, SMV has stricter requirements of approximately 2 ms. Encryption is not feasible for SMV, and management of encryption between MU and IED doesn't seem realistically possible, and that's often the cost of having wire-speed, reliable, real-time communications at high performance with very low latency.

An attacker can passively obtain all the SMV messages and spoof or replay messages. MITM is a big threat because it will cause significant latency to alter the communications, and provide incorrect IED readings.

DoS attacks are possible in both GOOSE and SMV; an attacker can overload subscribers with a load of messages.

VLAN is necessary for the performance and prioritizing communications of SMV, same as for GOOSE, and it might act as a security control to isolate SMV communications from attackers' access.

In later chapters of Cybersecurity Controls, we will also address more protections for all IEC 61850 communications protocols.

IEC 61850 Time Synchronization Requirements

How can you identify the latency or delays without having a time reference? In terms of time requirements, it's mandatory to have a reference that we can measure based on.

Following the discussion about different protocols, we identified that they have time requirements, and the accepted latency for some could be very low, approximately 2 ms.

Some measurements rely on time reference, such as the **Synchrophasor, a** device used to measure electrical quantity (voltage or current) based on time reference. Using protocols such as the Precision Time Protocol (PTP) can measure multiple points on the grid based on a unified and precise time reference.

So what are the time requirements in IEC 61850?

We have five classes of time synchronization in IEC 61850:

Class	Accuracy	Usage
T1	+/- 1 ms	Event logging
T2	+/- 100 µs	Zero Crossing
T3	+/- 25 µs	Protection
T4	+/- 4 µs	
T5	+/- 1 µs	

- µs: Microsecond
- ms: Millisecond

Each function has specific time requirements. Some protection functionalities have really low requirements; time synchronization is essential, and if it's not done accurately amongst devices, it will cause severe problems.

In an earlier chapter, we discussed the three different time protocols (SNTP, NTP, and PTP). We need to realize that the demand for accurate and precision time synchronization is vital to protection functionalities and reliable protocol communications.

Protection mechanisms will isolate the faulty parts *quickly* without disturbing the rest of the power grid systems. We have mentioned some protection components earlier, and usually, they will be used to take corrective actions to prevent the fault from impacting other parts of the system. And all these components must be accurately synchronized with the master clock.

Precision Time Protocol

An IEEE 1588 standard protocol, as the name indicates, it provides precision time synchronization; the accuracy in PTP synchronization is up to a nanosecond; and it uses hierarchal master-slave architecture for time synchronization.

The main components to be found are:

- Grandmaster Clock (Master) provides the central time reference and usually relies on a source for obtaining the time, such as GPS.
- Ordinary clock (Slave) obtains time synchronization from master clock.
- Bridge clocks (boundary and transparent clocks), an interface between master and slave to serve a specific subset of the network.

In IEC 61850, devices from different levels (Routers, Switches, IEDs, Servers) will need to synchronize using PTP as an Ordinary Clock (OC) with a Grandmaster Clock (GMC), which in turn will obtain the accurate time through another source (ex: GPS).

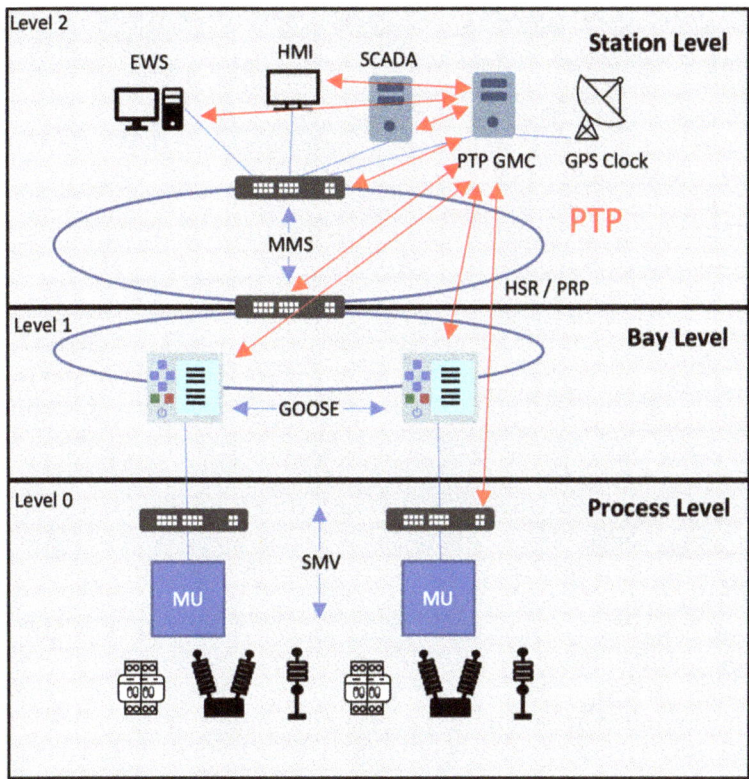

PTP Hierarchy

PTP works in the Master-Slave hierarchy; the Grand Master Clock (GMS) is the master that obtains the time from an external source and provides it to slaves, either directly to the Ordinary Clock (OC) or through a transparent clock to OC or through boundary clock to multiple OC.

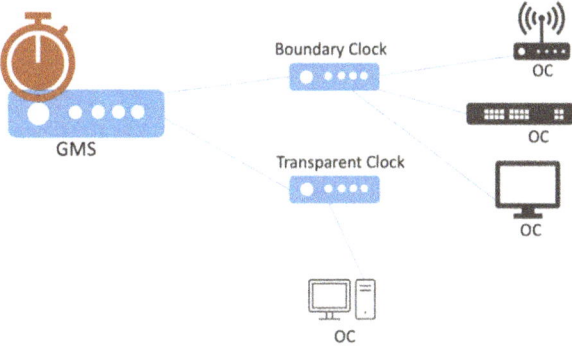

How does PTP work?

It makes some calculations based on the difference between when a message "Time-Sync" is sent and when it's received and continuously repeated; the slave clock then will adjust based on this difference.

In case there were bridge clocks between master and slave. Still, the difference is calculated based on the delay introduced through every clock from master to slave, and it will be the reference for adjustment.

When the target is to achieve synchronization accuracy up to the nanosecond, we have to consider some issues.

There is a delay between when GPS obtains the time and provides it to master clock then sends it to slave clock; in the following example, we see 0.01 seconds delay, then the time is taken to be sent to slave clock consumed 0.01 seconds. There are two delays: the processing on the master side until sending and the transmission delay from master to slave.

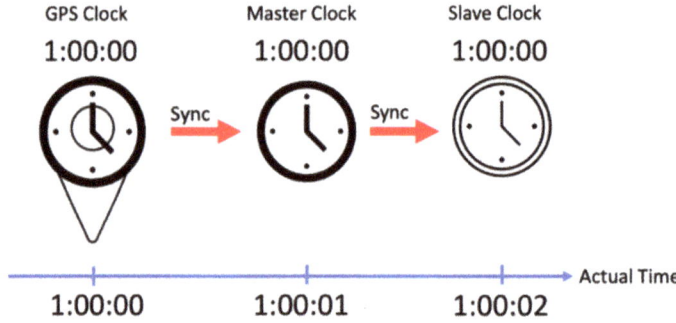

PTP protocol solves these issues in the following mechanism:

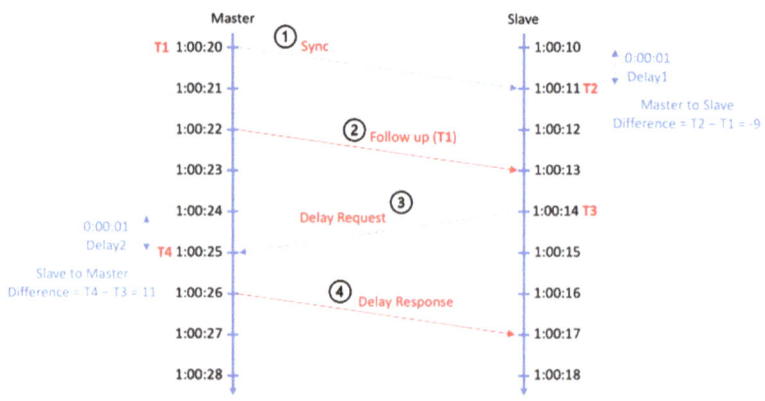

1) Master sends the Sync message with its timestamp (T1 = 1:00:20), after transmission *delay1* of (0:00:01) it's received by the slave at its timestamp (T2 = 1:00:11).
2) Master sends a follow-up message indicating the (T1).
3) Slave sends Delay request with its timestamp (T3 = 1:00:14), after transmission *Delay2* of (0:00:01) it's received by the master at its timestamp (T4 = 1:00:25).
4) Master sends delay response with a timestamp (T4 = 1:00:25) of when the delay request received.

Master to slave difference = T2 − T1 = -0:00:09
Slave to master difference = T4 − T3 = 0:00:11
One way delay = (Master to Slave difference + Slave to Master difference) / 2

$$= (-0:00:09 + 0:00:11) / 2$$
$$= (0:00:02)/2 = 0:00:01$$

Offset = Master to slave difference − one-way delay

$$= -0:00:09 - 0:00:01 = -0:00:10$$

So the slave clock is **less** by (**0:00:10**)

PTP uses UDP as a transport protocol and usually uses the following UDP ports:

- UDP 320 for general messages:
    - Announcement
    - Sync Follow up
    - Delay Response
    - P Delay Response Follow up
    - Management
    - Signaling
- 319 for event messages:
    - Sync
    - Delay Request
    - P Delay Request

o   P Delay Response

Let's take a look at the PTP capture (from github.com/freecores/):
General message: Announcement on UDP 320:

```
User Datagram Protocol, Src Port: 320, Dst Port: 320
Precision Time Protocol (IEEE1588)
    0001 .... = majorSdoId: gPTP Domain (0x1)
    .... 1011 = messageType: Announce Message (0xb)
    0000 .... = minorVersionPTP: 0
    .... 0010 = versionPTP: 2
    messageLength: 64
    domainNumber: 0
    minorSdoId: 0
    flags: 0x000c
    correctionField: 0.000000 nanoseconds
        correction: Ns: 0 nanoseconds
        correctionSubNs: 0 nanoseconds
    messageTypeSpecific: 0
    ClockIdentity: 0x008063ffff0009ba
    SourcePortID: 1
    sequenceId: 58
    controlField: Other Message (5)
    logMessagePeriod: 1
    originTimestamp (seconds): 1169232218
    originTimestamp (nanoseconds): 175326816
    originCurrentUTCOffset: 0
    priority1: 96
    grandmasterClockClass: 0
    grandmasterClockAccuracy: Unknown (0x00)
    grandmasterClockVariance: 128
    priority2: 99
    grandmasterClockIdentity: 0xffff0009baf82100
    localStepsRemoved: 128
    TimeSource: Unknown (0x80)
```

The announce message declares the capabilities of the clock; this is useful for establishing the Master-Slave hierarchy.

It will contain fields such as:

Message Type: Announce message

VersionPTP: PTP v 2

domainNumber: Logical grouping of clocks in one domain; if a message belongs to the domain, it will specify the number.

Correction: in a nanosecond, it indicates the two types of delays, obtaining the time from GPS and the transmission delay.

ClockIdentity: Unique identifier of the clock.

SequenceID: Sequence number for individual messages.

Event message: Sync on port UDP 319

Master clock could work in one step or two steps (adding follow-up); the time will be included if one step. If there are two steps, the time will be zero, and in the follow-up, time is provided.

```
User Datagram Protocol, Src Port: 319, Dst Port: 319
Precision Time Protocol (IEEE1588)
   0001 .... = majorSdoId: gPTP Domain (0x1)
   .... 0000 = messageType: Sync Message (0x0)
   0000 .... = minorVersionPTP: 0
   .... 0010 = versionPTP: 2
   messageLength: 44
   domainNumber: 0
   minorSdoId: 0
   flags: 0x0000
   correctionField: 0.000000 nanoseconds
      correction: Ns: 0 nanoseconds
      correctionSubNs: 0 nanoseconds
   messageTypeSpecific: 0
   ClockIdentity: 0x008063ffff0009ba
   SourcePortID: 1
   sequenceId: 115
   controlField: Sync Message (0)
   logMessagePeriod: 0
   originTimestamp (seconds): 1169232217
   originTimestamp (nanoseconds): 174421797
```

PTP Security

In IEC 61850, PTP plays a significant role in time synchronization. You are now aware of the synchronization requirements and timing accuracy important in the substations; the challenge is a minor change that could impact since the process and Bbay levels time requirements are very low.

PTP is subjected to MiTM, DoS, and injection attacks, and it's crucial to add security. There is a proposal for PTP v2 to add authentication and encryption, but it seems dropped and obsolete; it's essential to keep track of when PTP v2.1 will be released, including security.

Until the release and update of PTP to version 2.1, make sure to take the following steps:

- Redundancy is essential; you need to have redundant GMC and ensure the availability of redundant paths.
- Place GMC in DMZ.
- If the communication is over WAN or MPLS, add L2 or L3 encryption for PTP communications.
- Slaves can be configured to use NTP as a redundant protocol in PTP failure.
- Cybersecurity controls will be discussed in later chapters.

Chapter 10: Oil and Gas (O&G)

O&G are natural resources, and they form the primary sector of the energy industry. They directly impact the economy and our lives. O&G involves many complex processes and different phases from exploration until delivered as a consumer product.

O&G are usually grouped because they have similar phases and processes. Oil is still the dominant market, but since gas produces lower emissions than oil, it's being preferred by some countries and pushed onto the market through regulations.

The O&G operations are divided into three stages:

- Upstream
- Midstream
- Downstream

Upstream

It is the process of **Exploration and Production (E&P)**; this stage involves searching and exploration for crude oil and natural gas resources. Then it requires drilling and extraction.

The resources must be viable and have economic quantity and quality; during the E&P stage, once there are enough data about the potential reserve, drilling test wells will take place to extract samples to confirm quantity and quality.

Once it's confirmed as a viable resource, complete construction of the platform will begin, and it could be onshore or offshore.

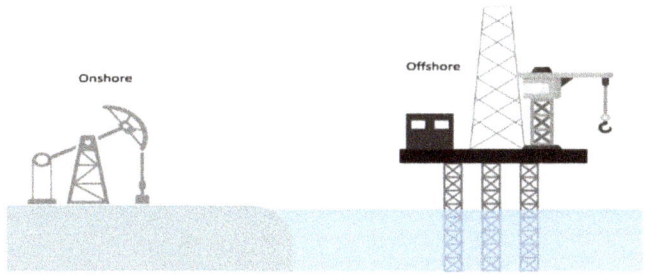

There are different vertical and horizontal drilling technologies, depending on the resource location and the best method to extract.

Such operations can be monitored and controlled remotely; SCADA systems are widely used to have a complete overview of the drilling operations.

The extracted O&G might be stored or shipped in different ways to refineries.

Midstream

The midstream stage involves transporting, storing, and wholesale marketing crude or refined petroleum products.

O&G will be transported by tanker ships and pipelines and stored in special tanks.

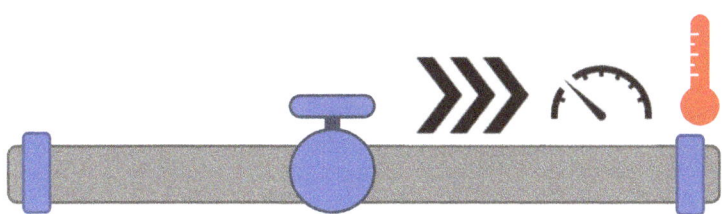

O&G Pipeline is one fundamental example of OT. SCADA systems monitor the pipelines and storage points; these systems will have monitoring and control over pumps, metering, flow, pressure, and storage measurements.

Over long distances from the pipeline, it's impossible to have the usual internet connectivity for SCADA; special equipment can be used with essential specifications. 5G will play a significant role in the near future for pipeline SCADA; a secure gateway with 5G capabilities is better for connecting to RTUs that monitor and control the sensors and actuators of the pipelines.

The power source is another challenge, the use of solar power and batteries for these devices is essential, but it also implies the need to have low power consumption.

Downstream

The downstream sector is refining petroleum crude oil, processing and purifying raw natural gas, and marketing and distributing products derived from crude oil and natural gas.

The downstream sector reaches consumers through products such as:

- Gasoline or petrol
- Kerosene
- Jet fuel
- Diesel oil
- Heating oil
- Fuel oils
- Lubricants
- Waxes
- Asphalt
- Natural gas
- Liquefied petroleum gas (LPG)
- Naphtha
- Hundreds of petrochemicals.

Midstream operations are often included in the downstream category and considered part of the downstream sector.

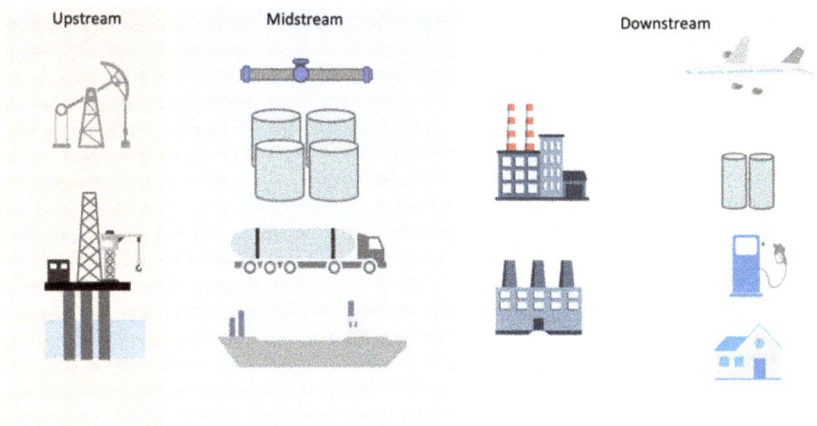

In each sector, various types of operations involve DCS, SCADA, and the use of IIoT.

This chapter will focus on the SCADA example of how the pipelines are monitored and controlled by using the SCADA system.

The importance of monitoring pipelines

Cyberattacks are considered one of the most significant threats to the O&G industry today. The pipeline is one of the targeted sectors, and the ransomware attack on the US Colonial Pipeline in 2021 was an eye-opener to the impact of such cyberattacks.

The pipeline is also a target of terrorist attacks; many attacks were reported in different countries; the pipeline is a very long weak link that is several thousand Km long; it requires physical security similar to countries' borders.

You must realize that pipeline is also a major political topic that could lead to conflict where some O&G producer countries tend to dominate the market. They could take an act of war to prevent other supply pipelines from reaching their market, and this could be under cover of terrorist and cyberattacks that might be escalated.

The pipeline can suffer from other problems, such as leakage, corrosion, and cracks, due to environmental effects or bad quality components or fixing. As you may realize, this requires proper monitoring along the pipeline.

Monitoring is required to detect any issue, and in case something is spotted, action must be taken. If there was a leakage detection at some point, the valves before and after the leakage must be closed, and pressure pumps must be stopped. That will follow physical inspection and maintenance.

How to detect an issue in the long pipeline?

The pipeline will have many control points that can stop the flow and terminate the pressure; all these points must be monitored and controlled; this implies the need for a large number of sensors and actuators.

Sensors will be used to measure:

- Pressure
- Flow rate
- Temperature

Various factors require different types of sensors. The oil density and viscosity are different from the gas density and viscosity, which means we need different sensors to measure flow rate. Other factors like the pipes' internal roughness and diameter are to be considered to measure the pressure.

A drop of pressure or flow rate at some point will indicate an issue that can scale from leakage to a total cut in the pipeline, and this involves many calculations because it's not a simple indicator; for example, the flow at the end of the pipeline feeding a storage tank will reduce pressure, the use of pumps will add pressure, and some pumps might suffer from problems that could lead to lower generated pressure, so the drop in pressure doesn't always mean leakage, and the use of these calculations and equations are often providing enough insights to categorize the problem and accordingly to take the appropriate action. As a result, different types of pressures must be calculated, such as the overall, static, and differential pressures, and it's taken from different points because the variation of readings from other points is another indicator.

Similarly, an increase of pressure and a drop in flow rate at one point where there is low pressure and a low flow rate at another point can be indicators of blockage.

The SCADA systems monitoring pipeline will have all required calculations applied to the continuous readings from the field and will alert or take actions according to the programmed logic.

What would be the impact if the SCADA systems lost connectivity with field devices?

As indicated in the SCADA chapter, the system will continue to work in the field, and once the connectivity is back, it will resume monitoring and control, but during the downtime of the link, the control room is in the dark and has no clue of what is happening in the field, so one of the concerns is to have an attack on the connectivity first to isolate the monitoring capabilities and then to cause a physical impact on the pipeline. This will increase the time to respond and will give attackers more time to cause damage.

Infrastructure:

SCADA, as discussed earlier, relies on long-distance and low-bandwidth communications; we have a SCADA server (Master Terminal Unit, MTU) in the Control Room and Remote Terminal Units (RTU) in the field connecting the field devices.

Due to the distance between different sensors and actuators, wireless-capable RTU and field devices might be used, and other field devices might be terminated directly to the RTU.

On the other hand, RTU communicates over WAN links with the SCADA server in the control room; these WAN links could be 4G/5G, copper/fiber, or satellite links.

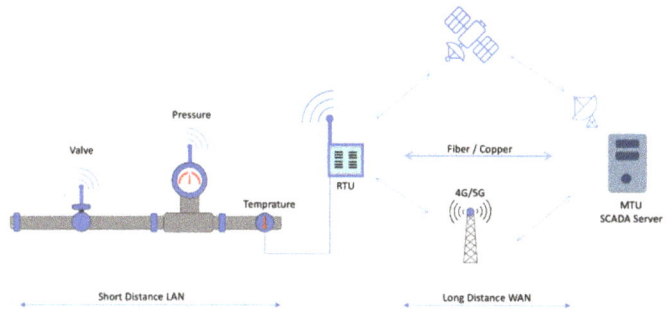

Communication Protocols

There's no one protocol used as the standard for O&G pipeline monitoring communications. Most protocols can fit when we have low bandwidth and speed requirements while carrying simple analog and digital values.

We will consider the Modbus protocol because it's widely used; it's the basic protocol that anyone in the OT world should be aware of. This chapter will discuss it since it applies to the O&G pipeline.

Modbus

Modicon invented the PLC controller and Modbus protocol in 1979. Modicon changed the shape of the OT market by introducing PLC that can be programmed to do any function and a simple communication protocol that can carry I/O signals of analog and digital values. Modicon was later acquired by Schneider Electric and became part of it.

Modbus communicates simple raw messages without restrictions, without authentication, and uses a simple header. It communicates in simple Master/Slave requests and replies.

It's a layer 7 application protocol that uses three types of PDU (protocol data unit):

1) Request
2) Response
3) Exception Response (in case of error)

At first, the protocol didn't use Ethernet. It was designed to communicate over serial links. The first version used ASCII as data communications, where the representation of data using ASCII caused a large message size.

The communicated message is represented by ASCII code; each 1 byte is represented using two ASCII characters.

Example:

A byte of value: *0xA9* will be decoded as 0x41 and 0x39 because *A* = 0x41 and *9* = 0x39. Well, it's not efficient to have one byte represented in two ASCII characters, so naturally, this method needed to evolve.

The messages are communicated over the serial link; each byte will be sent from the least significant bit from the left, bit by bit to the right, until the most significant bit on the right.

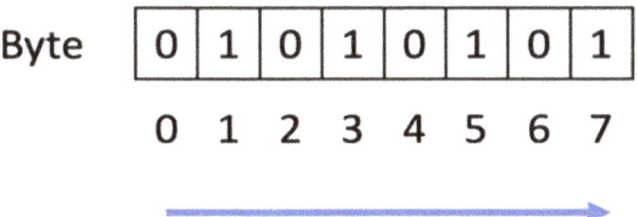

Modbus ASCII

- Colon ':' to indicate the beginning of a frame.
- ASCII characters (large messages)
- LRC for error checking (not reliable)

Modbus RTU is different and incompatible with Modbus ASCII

- Hexadecimal bytes data representation, better than ASCII
- CRC for error checking, and it's more reliable than ASCII
- A quiet time to indicate the end of transmission

How are the messages formed?
The message will contain the following fields:

- Address: slave address.
- Function Code: Each operation is requested by the function code.
- Data: data of the function code, either request or response.
- CRC or LRC: error handling.

The use of serial links (RS-232, RS-422, RS485) can vary in terms of supported length and speed; the first stage of Ethernet communication was to encapsulate Modbus RTU over TCP segments. Then there was the introduction of Modbus TCP, which is different and incompatible with Modbus RTU over TCP.

Today Modbus TCP is widely adapted; it's compatible with almost all systems due to its simplicity and efficiency in transferring texts and numbers in various formats, and it's common to be used across different verticals in operational technology.

Modbus TCP

Modbus TCP communicates in a Master/Slave method, where the master is the client and the slave is the server. Slave has a set of addresses, where each address represents an I/O device.

When the master device wants to read/write to an I/O device, it sends a request and address; in return, the slave will answer with the address associated with the response, which could indicate an error due to a wrong address or other reasons.

Let's take an example of Modbus communication:

Two communicating devices:

- Client: HMI
- Server: PLC

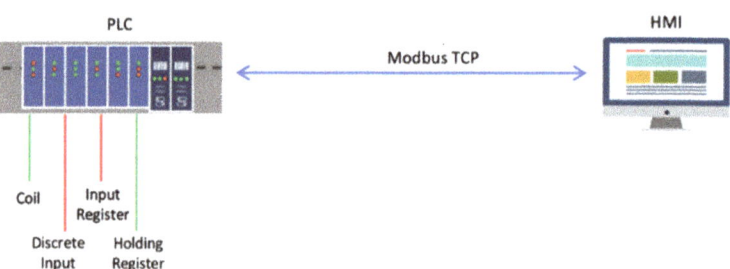

In Modbus TCP, the PLC is identified by IP address; there's also the use of 'Unit ID,' which is an identifier of Modbus PLC; depending on the implementation, there might be a single Unit ID or multiple on the same PLC (same IP address), and also when using Modbus bridges that converts Modbus TCP to other serial implementation of Modbus, for that reason it's not enough to use IP Address. Still, you need to specify which unit ID.

Unit ID is a numeric number in the range of 1–255.

The I/O devices on the PLC are four types:

- Coil (Read/Write): Digital 1/0
- Discrete Input (Read-Only): Digital 1/9
- Input Register (Read-Only): Numeric 16-bit
- Holding Register (Read Write): Numeric 16-bit

Each one of the I/O has specific methods (functions) to access for read or write; when you have Read-Only I/O, it means you can only read from it; usually, they are connected to sensors to get readings from, and in the case of Read/Write then it's connected to an output device where you can write values to.

Each I/O device, when accessed through a specific function, needs to select the address of the tag/set point and where you can find up to 9999 possible addresses.

Each I/O type is identified by a number followed by available addresses as follows:

- Coil: 00001 – 09999
- Discrete Input: 10001 – 19999
- Input Register: 30001 – 39999
- Holding Register: 40001 – 49999
- Why starting from 0001 and not 0000 depends on the vendor implementation, and it's essential to know this information when programming Modbus because, according to the vendor, it could be Zero's addressing (uncommon) or one's address as in the above list.
- For registers, where they have two bytes' values (16 bit) also, it's crucial to know-how is the vendor implementation, as some vendors use different orders of transfer, which may transfer 1st byte followed by 2nd byte, or it's possible to transfer 2nd byte then 1st byte.
- In some cases, when the registers are required to host long values (more than 2 bytes), they can store several addresses for a single value, which can be accessed by providing the range of addresses that host the desired value. Still, then the previous point is essential to know which byte is being transferred first.

Suppose you need to read the state of the device connected to Discrete Input; you need to send the following information within a request:

- Specify the IP address of the PLC
- The unit ID
- Use function code 0x02, which is used to 'read discrete value'
- Provide the address of the discrete input

The response might be error 'exception error message' due to wrong of any value IP/Unit ID/Function Code/address, or it will provide the requested reading, including all the data used in the request.

What are the common function codes of Modbus?

Code	Description
01	Read coil status
02	Read discrete input status
03	Read holding register
04	Read input register
05	Write single coil
06	Write single register
15	Write multiple coils
16	Write multiple registers
23	Read/Write multiple registers

- When accessing a single value, only the value's address is required, but when accessing a value stored in multiple addresses, the start address and length are needed.

Suppose you need to write a set point to the holding register, and it's stored in three addresses (0100, 0101, and 0102). To access the holding register address begins with '4,' and you will provide the address of the first register, '40100,' and the length of '3.'

How is the Modbus TCP communication payload formed?

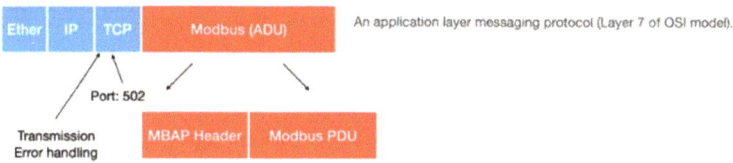

Modbus (ADU): Application Data Unit (256 bytes)
MBAP: Modbus Application Protocol (7 bytes)
PDU: Protocol Data Unit (up to 249 bytes)

- Modbus RTU is limited to 256 bytes of payload, and Modbus inherited the same limitation.

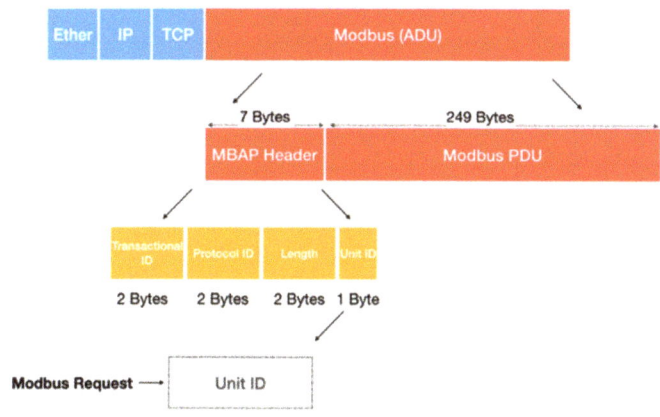

The MBAP header will include the following

- Transactional ID: an ID used to synchronize between messages request/response
- Protocol ID: Reserved and always '0'
- Length: Number of remaining bytes
- Unit ID: the identifier of the PLC

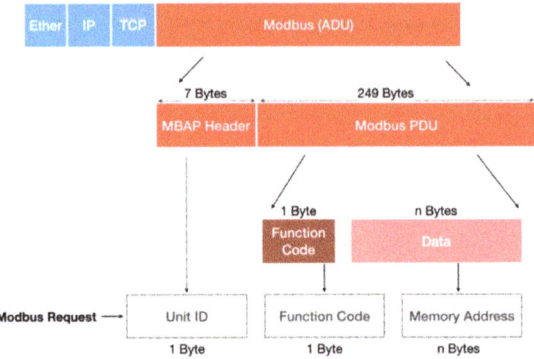

Modbus PDU contains the function code and memory address, and in case of response, it will include the data.

Let's take the following example:

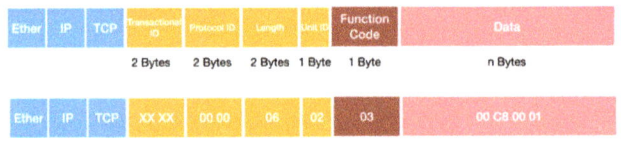

- XX XX is the transactional ID
- 00 00 Protocol ID (reserved)
- Six bytes length
- 02 unit ID
- 03 Function Code (Read holding register)
- 00 C8: The start address (200 in decimal which will begin by '4' for holding register 40200)
- 00 01: Number of address(es) to read (single address)

Now let's see some examples of Modbus communication:
Request: Read Coil

```
Internet Protocol Version 4, Src: 192.168.0.134, Dst: 172.17.0.2
Transmission Control Protocol, Src Port: 49744, Dst Port: 502, Seq: 1, Ack: 1, Len: 12
Modbus/TCP
    Transaction Identifier: 40968
    Protocol Identifier: 0
    Length: 6
    Unit Identifier: 1
Modbus
    .000 0001 = Function Code: Read Coils (1)
    Reference Number: 0
    Bit Count: 100
```

Response:

```
Internet Protocol Version 4, Src: 172.17.0.2, Dst: 192.168.0.134
Transmission Control Protocol, Src Port: 502, Dst Port: 49744, Seq: 1, Ack: 13, Len: 22
Modbus/TCP
    Transaction Identifier: 40968
    Protocol Identifier: 0
    Length: 16
    Unit Identifier: 1
Modbus
    .000 0001 = Function Code: Read Coils (1)
    [Request Frame: 7]
    [Time from request: 0.000378000 seconds]
    Byte Count: 13
    Bit 0 : 1
    Bit 1 : 1
    Bit 2 : 1
    Bit 3 : 1
    Bit 4 : 1
    Bit 5 : 1
    Bit 6 : 1
    Bit 7 : 1
    Bit 8 : 1
    Bit 9 : 1
    Bit 10 : 1
    Bit 11 : 1
    Bit 12 : 1
    Bit 13 : 1
    Bit 14 : 1
    Bit 15 : 1
```

Note the following:

- Transaction identifier is the same (40968) for request and response, so they are linked.
- Unit ID: is '1.'
- Function Code: '1' Read Coils.
- In Request: Reference is number is '0,' which means it requested to read from Address 00000 and Bit count is 100, so it will read all values from 00000–00099.
- In Response, it provided all the addresses from bit 0 (00000) until the end with the associated value; in this case, it was always '1' or 'logical true.'

Another example:

Request: Write Single Register

```
Internet Protocol Version 4, Src: 192.168.0.134, Dst: 172.17.0.2
Transmission Control Protocol, Src Port: 49744, Dst Port: 502, Seq: 529, Ack: 1273, Len: 12
Modbus/TCP
    Transaction Identifier: 52232
    Protocol Identifier: 0
    Length: 6
    Unit Identifier: 1
- Modbus
    .000 0110 = Function Code: Write Single Register (6)
    Reference Number: 0
    Data: 000b
```

Write the value: '000b' to register reference 0 (40000) using function code: '6'

Response:

```
Internet Protocol Version 4, Src: 172.17.0.2, Dst: 192.168.0.134
Transmission Control Protocol, Src Port: 502, Dst Port: 49744, Seq: 1273, Ack: 541, Len: 12
Modbus/TCP
    Transaction Identifier: 52232
    Protocol Identifier: 0
    Length: 6
    Unit Identifier: 1
- Modbus
    .000 0110 = Function Code: Write Single Register (6)
    [Request Frame: 144]
    [Time from request: 0.000298000 seconds]
    Reference Number: 0
    Data: 000b
```

Confirmation of data written to the holding register.

Modbus protocol is considered a simple protocol for communicating different types of tags and set points, which explains the wide usage not only in O&G but also in many other industries and operations.

Modbus Security

You may have already noticed that there are no means of encryption or authentication, making it a fragile protocol.

I have never seen any implementation in the production of TLS security for Modbus. However, it's possible, and I firmly believe that Modbus is an easy choice for simple communication and applications. The TLS complexity wouldn't be a good fit for such types.

Assets communicating over Modbus should be appropriately monitored by security controls and should be entirely isolated. It's similar to other non-secure protocols, but one additional risk is that it's limited and straightforward. You can find various tools that can communicate over Modbus and if attackers have access to the IP network where assets using Modbus can perform any damage.

In the Cybersecurity Controls chapter, we will discuss how to mitigate the risk of such insecure protocols.

Chapter 11: Building Management Systems (BMS)

Also referred to as Building Automation Systems (BAS).

A type of operational technology (not industrial in nature) that manages, monitors, and automates a building's equipment.

Examples of BMS:

- Air Conditioning
- Heating
- Ventilation
- Lighting
- Safety
- Plumping
- Fire Alarm
- Power
- Water
- Gas
- And various IoT sensors

It can also integrate with other systems in the building, such as access doors and CCTV cameras.

The ideal objective of BMS is to centralize the control and monitoring of all types of equipment in the building and function as a unified, integrated system. This makes it easier to monitor and control and, hence, more efficient; it can also help reduce the cost and power consumption; scheduling different systems to work at a particular time would maintain a sustainable environment and reduce cost by eliminating unnecessary run times.

One use case of multitenancy and facility management is providing a better experience for the tenant while being able to e-activate the services and reduce the usual lengthy procedures.

On large campuses or stadiums, the BMS plays a significant role in monitoring and controlling large areas; imagine if operators need to use a different system each time they need to change the state of access doors or control the lighting!

In BMS, however, there are many different OEM vendors of equipment with an explosive number of IoT devices that can be used for various functionalities. This is one of the main challenges of BMS: To monitor and manage these multiple types of equipment.

How about an example of luxury facilities: a steam room, warm/cool swimming pool, Jacuzzi, or meeting room facilities? They have different types of monitoring and control; they need to provide a safe and improve the user experience at the same time.

In BMS, if you have noticed, it's about Cyber-Physical systems that directly serve consumers; it's not like other operational technologies where trained and knowledgeable teams use them; in this case, safety and ease of use are essential, and this is important enough to make sure they are functioning securely from cyberattack.

Facilities management technical teams also use BMS to monitor the health and performance of the systems; as it's similar to many OT systems, BMS uses sensors to monitor variables such as heat and pressure for parts of the systems.

In BMS, we will take the BACnet Protocol as the most commonly used protocol. The key to BACnet/IP is that any device that supports it can communicate with other devices using the same protocol. Hence, it achieves **interoperability** across different vendors.

BACnet

A data communication protocol for **B**uilding **A**utomation and **C**ontrol **N**etworks.

It's developed under the auspices of the American Society of Heating, Refrigerating, and Air-Conditioning Engineers (ASHRAE).

BACnet is an American national standard, a European standard, a national standard in more than 30 countries, and an ISO global standard.

BACnet provides standards for interoperability:

- Data sharing
- Alarm and event management
- Trending
- Scheduling
- Remote device and network management

BACnet applications:

- HVAC control
- Fire detection and alarm
- Lighting control
- Security
- "Smart" elevators
- Utility company interface

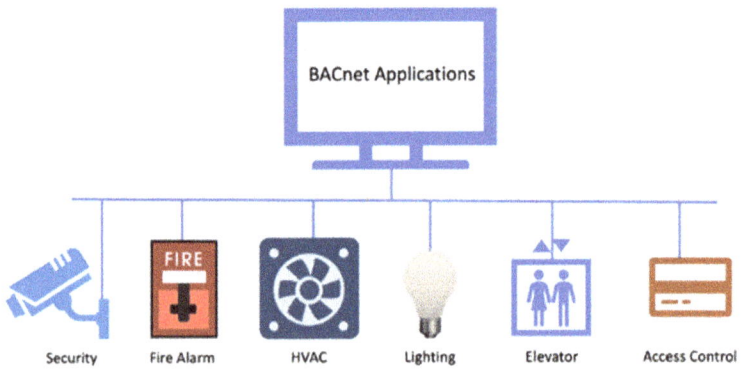

Why BACnet?

Traditionally, buildings had various systems; each has a particular purpose; they are isolated and communicated over proprietary protocols. Each will require a separate operator and cannot integrate except through expensive solutions.

Due to vendor-lock, it's also limiting customers to only purchase solutions from the same vendor. As a result, customers had to invest more in systems and hire more operators while getting fewer features and capabilities.

When BACnet was developed as an open-vendor protocol, it opened all possibilities to have unified monitoring, and a single operator could manage the system. Nevertheless, customers are free to choose from the right solution and get better features and capabilities.

How does the protocol work?

By introducing the concept of object, to simplify the access to a particular function, where an object is a collection of information with a unique identifier that can be accessed in standard methods.

In essence, the object represents a physical I/O or a software process; each has a set of attributes (properties). The I/O would have values (digital and analog), where a process has set points (ex: time, temperature).

Originally, BACnet defined 18 standard object types, but with evolving applications of the protocol and OEM vendors' adaption, there are more than 60 types today, and it's growing since OEM vendors can define their custom object types, making it flexible.

List of standard objects

OBJECT	EXAMPLE OF USE
Analog Input	Sensor input
Analog Output	Control output
Analog Value	Set point or other analog control system parameter
Binary Input	Switch input
Binary Output	Relay output
Binary Value	Binary (digital) control system parameter
Calendar	Defines a list of dates, such as holidays or special events, for scheduling.
Command	Writes multiple values to multiple objects in multiple devices to accomplish a specific purpose, such as day-mode to night-mode or emergency mode.
Device	Properties tell what objects and services the device supports and other device-specific information such as vendor, firmware revision, etc.
Event Enrollment	Describes an event that might be an error condition (e.g., "Input out of range") or an alarm that other devices to know about. It can directly tell one device or use a Notification Class object to tell multiple devices.
File	Allows read and write access to data files supported by the device.

Group	Provides access to multiple properties of multiple objects in a read single operation.
Loop	Provides standardized access to a "control loop."
Multi-state Input	Represents the status of a multiple-state process, such as a refrigerator's On, Off, and Defrost cycles.
Multi-state Output	Represents the desired state of a multiple-state process (such as It's time to cool, It's cold enough and it's time to defrost).
Notification Class	Contains a list of devices to be informed if an event enrollment object determines that a warning or alarm message needs to be sent.
Program	Allows a program running in the device to be started, stopped, loaded, and unloaded, and reports the present status of the program.
Schedule	Defines a weekly schedule of operations (performed by writing to a specified list of objects with exceptions such as holidays. Can use a calendar object for the exceptions.

Each object has a set of possible properties, including a unique identification of the object and properties describing the object and the object's current state, where you can see the property in pairs of name and value.

Example of Analog Input Object Properties:

PROPERTY	BACnt	EXAMPLE
Object_Identifer	Required	Analog Input #1
Object_Name	Required	"AI 01"
Object_Type	Required	Analog Input
Present_Value	Required	68.0
Description	Optional	"Outside Air Temperature"
Device_Type	Optional	"10k Thermistor"
Status_Flags	Required	In_Alarm, Fault, Overridden, Out_Of_Service flags
Event_State	Required	Normal (plus various problem-reporting states)
Reliability	Optional	No_Fault_Detected (plus various fault conditions)
Out_Of_Service	Required	False

Update_Interval	Optional	1.00 (seconds)
Units	Required	Degrees-Fahrenheit
Min_Pres Value	Optional	-100.0, minimum reliably read value
Max_Pres Value	Optional	+300.0, maximum reliably read value
Resolution	Optional	0.1
COV_Increment	Optional	Notify if Present_Value changes by increment: 0.5
Time Delay	Optional	Seconds to wait before detecting out-of-range: 5
Notification Class	Optional	Send COV notification to Notification Class Object: 2
High_Limit	Optional	+215.0, Upper normal range
Low Limit	Optional	-45.0, Lower normal range
Dead band	Optional	0.1
Limit_Enable	Optional	Enable High-limit-reporting, Low-limit-reporting.
Event_Enable	Optional	Enable To_Offnormal, To_Fault, To_Normal change reporting.
Acked_Transitions	Optional	Flags indicating received acknowledgments for above changes.
Notify_Type	Optional	Events or Alarms

Service Request/Reply

The physical devices or software processes are now represented by objects, where each has its list of properties and identifiers. The BACnet protocol offers 32 services to access these objects/properties, and it's the method used to monitor and control the BMS in BACnet.

There are five categories of Services:

1. **Object Access Services** to access and manipulate the properties of BACnet objects;
2. **Alarm and Event Services** to obtain alarms, events, and change of value notifications;

3. **File Access Services** to access and manipulate files contained in BACnet devices;
4. **Remote Device Management Services** for device discovery, time synchronization, device communication control, and reinitialization;
5. **Virtual Terminal Services** to facilitate the bi-directional exchange of character-oriented data.

Furthermore, the services can be confirmed (expecting a reply) or unconfirmed (no reply expected).

BACnet Network

Since the protocol was developed in 1987, it has evolved over different media, which resulted in various networks and ways of communication.

BACnet over MS/TP

Master – Slave/Token Passing

This old technology used tokens to pass requests and responses, and the devices were either master (sending requests) or slave (submitting responses); the communication used to take place over serial links, which was low speed and not efficient compared to BACnet over IP.

BACnet over IP

The communications between devices using IP addresses are more flexible and efficient because IP is routable, so the concept of the BACnet router is introduced to route BACnet communication over different networks easily and securely because it can work over a secure VPN.

IP version of BACnet also adapted API, which facilitates communications amongst different systems.

There are other types of networks used in BACnet:

- Ethernet, ISO/IEC 8802-3, high-speed LAN but non-routable.
- ARCNET, ATA/ANSI 878.1, local communications, and dedicated communication integrated circuits (IC).
- LonTalk provides a basic mechanism for exchanging messages, regardless of the type of the messages. Although it became an open

protocol, it still has limitations, and only devices that support the LonTalk chip can be used for BACnet over LonTalk.
- Point to point PTP, its low speed and used to be the choice of dialup.

IP protocols are the new language spoken across all digital devices today. With the efficiency and speed, BACnet/IP is the most suitable protocol to be used today.

BACnet/IP communicates by continuously sending broadcast messages that are received by all devices in the network; this is an important technique to discover all devices in the network and share information. But broadcast might be an issue if there are multiple networks that all receive the broadcast messages. This could cause congestion and might bring networks down; for this reason, BACnet/IP uses a device called a **BBMD**.

BACnet Broadcast Management Device (BBMD)

In each small subnet, there must be a single BBMD device; this device manages broadcast within the subnet.

Suppose a message is sent to a device in another subnet. In that case, the BBMD in the sender subnet will send a unicast message to the BBMD in the destination subnet, which delivers it locally to the destination device.

This allows unicast communication between different subnets in large implementations and limits broadcasts within the same subnet.

It's important to know that there must be only one single BBMD in the same subnet (broadcast domain), or this will cause issues in the communications.

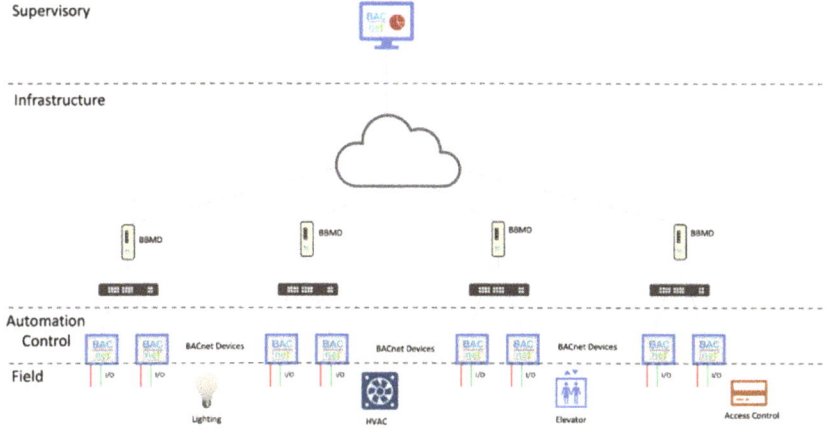

BACnet IP Ethernet Frame format:

BVLL: BACnet Virtual Link Layer

- It handles virtual links between BACnet devices and determines if the message will be broadcast or unicast.

NPDU: Network-Layer Protocol Data Unit

- Contains details on the network between BACnet devices, indicating the fields (ex: version), following the mapping over the network using IP from BACnet to BACnet device, it's mapped using the NPDU.

APDU: Application Layer Protocol Data Unit

- Depending on the service request/response and type of object, the payload of BACnet will be details around the data exchanged.

The Discovery

Using the broadcast message "Who is?" and devices claiming themselves using the broadcast message "I Am."

Broadcast: Who is?

```
> User Datagram Protocol, Src Port: 62881, Dst Port: 47808
∨ BACnet Virtual Link Control
    Type: BACnet/IP (Annex J) (0x81)
    Function: Original-Broadcast-NPDU (0x0b)
    BVLC-Length: 4 of 12 bytes BACnet packet length
∨ Building Automation and Control Network NPDU
    Version: 0x01 (ASHRAE 135-1995)
    > Control: 0x20, Destination Specifier
    Destination Network Address: 65535
    Destination MAC Layer Address Length: 0 indicates Broadcast on Destination Network
    Hop Count: 255
∨ Building Automation and Control Network APDU
    0001 .... = APDU Type: Unconfirmed-REQ (1)
    Unconfirmed Service Choice: who-Is (8)
```

You may notice the following:

- The BVLC "BACnet Virtual Link Control" is usually part of the BVLL.
- BVLC specifies the version and the function (original broadcast NPDU).
- NPDU tells the version, the control, the destination network, MAC address, and Hop count to avoid loops.
- APDU indicates the type: Unconfirmed Request '1' (doesn't expect an answer).
- The unconfirmed requested service is: Who is '8'?

Broadcast: I am

```
> User Datagram Protocol, Src Port: 47808, Dst Port: 47808
v BACnet Virtual Link Control
    Type: BACnet/IP (Annex J) (0x81)
    Function: Original-Broadcast-NPDU (0x0b)
    BVLC-Length: 4 of 25 bytes BACnet packet length
v Building Automation and Control Network NPDU
    Version: 0x01 (ASHRAE 135-1995)
    > Control: 0x20, Destination Specifier
    Destination Network Address: 65535
    Destination MAC Layer Address Length: 0 indicates Broadcast on Destination Network
    Hop Count: 255
v Building Automation and Control Network APDU
    0001 .... = APDU Type: Unconfirmed-REQ (1)
    Unconfirmed Service Choice: i-Am (0)
    > ObjectIdentifier: device, 1001
    > Maximum APDU Length Accepted: (Unsigned) 1476
    > Segmentation Supported: segmented-both (0)
    > Vendor ID: ProtoSense Technologies (723)
```

In the 'I am' Broadcast, the device announces itself with specifications; what is essential in the APDU is that it informs us of its unique object identifier "1001" and other properties and descriptions of the object.

If you want to monitor/control a device, it requires subscribing to the BACnet device; based on the "I Am" broadcast message, we know the details about the device, so in that case, it will be a unicast confirmed message because we need to subscribe to the specific device and we need confirmation.

Confirmed Request:

```
> User Datagram Protocol, Src Port: 62881, Dst Port: 47808
> BACnet Virtual Link Control
∨ Building Automation and Control Network NPDU
     Version: 0x01 (ASHRAE 135-1995)
   > Control: 0x04, Expecting Reply
∨ Building Automation and Control Network APDU
     0000 .... = APDU Type: Confirmed-REQ (0)
   > .... 0010 = PDU Flags: 0x2
     .111 .... = Max Response Segments accepted: Greater than 64 segments (7)
     .... 0101 = Size of Maximum ADPU accepted: Up to 1476 octets (fits in an ISO 8802-3 frame) (5)
     Invoke ID: 0
     Service Choice: readProperty (12)
   > ObjectIdentifier: device, 1001
   > Property Identifier: structured-object-list (209)
```

- In NPDU, it indicates it's a confirmed request and expecting a reply.
- Service '12': Read Property
- Object ID: device, 1001 as initially announced during the 'I am' broadcast message.
- Requesting structured object list '209'

If the requested property "structured object list" doesn't exist, the device will reply with the error code '32' Unknown property.

Then you may request another property, for example: '76' Object list, and the device will acknowledge if the property requested exists:

```
> User Datagram Protocol, Src Port: 47808, Dst Port: 62881
> BACnet Virtual Link Control
> Building Automation and Control Network NPDU
∨ Building Automation and Control Network APDU
     0011 .... = APDU Type: Complex-ACK (3)
   > .... 0000 = PDU Flags: 0x0
     Invoke ID: 1
     Service Choice: readProperty (12)
   > ObjectIdentifier: device, 1001
   > Property Identifier: object-list (76)
   > {[3]
   > ObjectIdentifier: device, 1001
   > ObjectIdentifier: notification-class, 1
   > ObjectIdentifier: binary-input, 0
   > }[3]
```

The response, in this case, came in the form of a complex ACK because it provides a list of objects, similar to the JSON format {"Device, 1001," "notification class 1," "binary input 0"}.

Now the client knows what properties to request, reading:

```
> User Datagram Protocol, Src Port: 62881, Dst Port: 47808
> BACnet Virtual Link Control
> Building Automation and Control Network NPDU
∨ Building Automation and Control Network APDU
    0000 .... = APDU Type: Confirmed-REQ (0)
    .... 0010 = PDU Flags: 0x2
    .111 .... = Max Response Segments accepted: Greater than 64 segments (7)
    .... 0101 = Size of Maximum ADPU accepted: Up to 1476 octets (fits in an ISO 8802-3 frame) (5)
    Invoke ID: 2
    Service Choice: readPropertyMultiple (14)
> ObjectIdentifier: device, 1001
∨ listOfPropertyReferences
    ∨ {[1]
        .... 1... = Tag Class: Context Specific Tag
        0001 .... = Context Tag Number: 1
        .... .110 = Named Tag: Opening Tag (6)
    ∨ Property Identifier: all (8)
        ∨ Context Tag: 0, Length/Value/Type: 1
            .... 1... = Tag Class: Context Specific Tag
            0000 .... = Context Tag Number: 0
            Length Value Type: 1
            Property Identifier: all (8)
    ∨ }[1]
        .... 1... = Tag Class: Context Specific Tag
        0001 .... = Context Tag Number: 1
        .... .111 = Named Tag: Closing Tag (7)
```

And in this case, it has requested to read all properties '8'; as a result, the device will respond by providing a complete list of all properties:

```
> User Datagram Protocol, Src Port: 47808, Dst Port: 62881
> BACnet Virtual Link Control
> Building Automation and Control Network NPDU
∨ Building Automation and Control Network APDU
    0011 .... = APDU Type: Complex-ACK (3)
  > .... 0000 = PDU Flags: 0x0
    Invoke ID: 2
    Service Choice: readPropertyMultiple (14)
> ObjectIdentifier: device, 1001
∨ listOfResults
  > {[1]
  > Property Identifier: object-name (77)
  > {[4]
  > Object Name
  > }[4]
  > Property Identifier: object-type (79)
  > {[4]
  > object-type:   device (8)
  > }[4]
  > Property Identifier: object-identifier (75)
  > {[4]
  > ObjectIdentifier: device, 1001
  > }[4]
  > Property Identifier: system-status (112)
  > {[4]
  > system-status:   operational (0)
  > }[4]
  > Property Identifier: vendor-name (121)
  > {[4]
  > vendor-name: UTF-8 'ProtoSense Technologies'
  > }[4]
  > Property Identifier: vendor-identifier (120)
  > {[4]
  > vendor-identifier: (Unsigned) 723
  > }[4]
  > Property Identifier: model-name (70)
  > {[4]
  > model-name: UTF-8 'Beamer1'
  > }[4]
  > Property Identifier: firmware-revision (44)
  > {[4]
  > firmware-revision: UTF-8 '1.2'
  > }[4]
  > Property Identifier: application-software-version (12)
  > {[4]
  > application-software-version: UTF-8 '1.2'
  > }[4]
  > Property Identifier: location (58)
  > {[4]
  > location: UTF-8 'Riyadh'
```

You will see that many properties have been shared on this object, and it is worth mentioning I have generated this communication using the BACnet simulator from ProtoSense Technologies.

The communication flows should make sense to you; they will differ based on the object type and service requested, but the flow will always be similar.

How about security?

BACnet IP has many security applications compared to other discussed protocols, which is why it is appreciated to see it implemented since BACnet devices are distributed on customer-facing premises. They are often exposed to physical access. Although physical access usually eliminates all other security aspects, at least altering messages or hijacking sessions would be challenging for attackers with such an implementation.

You always have to make sure that the protocol implements these security mechanisms because they are available and applicable.

BACnet relies on the use of **Shared Secret Keys**, where device and user **authentication** are achieved by using message signature and shared signature key.

Confidentiality of the messages is achieved by encrypting the payload and shared encryption keys. Although earlier we didn't give enough attention to confidentiality in other operational technologies, BMS may expose much information if the messages are read in clear text, giving attackers the chance to map the devices and identify the weakest entry points visually.

The shared secret keys are continuously distributed in pairs:

- Signature key
- Encryption key

There are six types of shared secret keys:

1) General Network Access Key, used for:
   a. Device and object binding
   b. Tunnel encryption
   c. User authentication from untrusted devices
- All devices must be given a General-Network-Access-Key, similar to the IT Network Access Control (NAC) authentication pre-connect, where the device must authenticate before accessing the network.

2) User Authenticated:
- It is distributed to client devices that are trusted to authenticate or to devices that don't have HMI.
3) The application-specific key is used for application-level authentication.
4) Installation keys are distributed temporarily to a small set of devices to access specific maintenance and configuration tasks.
5) Distribution keys are used to distribute the above four keys securely.
6) Device master keys, like Private keys in asymmetric key encryption, should not be transmitted over the wire.

Securing Messages

Can be achieved by encapsulating the payload inside a secure payload and using hashing algorithms (HMAC, MD5, SHA-256) to sign the message to mitigate the replay attack.

Message-ID and Timestamp are basic controls, but they can also help link appropriate request/response and the right timing of the conversation context.

BACnet Secure Connect (SC)

It was announced in 2020, so it's very recent, considering writing these words in April 2022, and it provides device authentication and message encryption that has been added to the protocol standard. And there will be a further enhancement for security down the road to include certificate management and authorizing interoperable messages.

BACnet, if not implemented securely, might be subjected to the following attacks:

- Session hijack
- Altering messages
- Rouge devices
- MAC Spoofing
- DoS
- L2/L3 network attacks

In summary, BACnet IP is more advanced in the adaptation of security controls due to its non-industrial nature and being closed to IT in many cases.

When you refer to BACnet.org, you will find plenty of security recommendations that haven't been discussed in this chapter, the subject is vast, at the end of the day your mission is to check if the BACnet communication is secure. The devices/users are also secured, and all security controls are implemented.

We have established beyond the basics of the BMS, BACnet, and security controls, which is the objective of this chapter; if you are working on a project to a secure BMS using the BACnet/IP protocol, you have the common knowledge to know what to look for, and in the later chapter of Cybersecurity Controls, you will also understand what other controls to be implemented to make sure of a safe and secure operation.

Part 4

Chapter 12: OT Cybersecurity Standards and Regulations

Cybersecurity has been made an essential part of OT. We have discussed many trends that led to the importance of cybersecurity for OT, such as Industry 4.0, an adaptation of TCP/IP communications, and various attacks on critical infrastructure.

There have been many initiatives from well-respected organizations to provide guidelines specifically for **IACS** systems to help them reduce the risk and operate safely and securely, such as CISA, NIST, and ISA. The latter community, ISA99, had previously published standards and best practices, but it was then found not efficient enough to ensure safety, integrity, reliability, and security following successful cyberattacks on IACS.

Then the International Society of Automation (ISA) and the International Electro technical Commission (IEC) Technical Committee 65/Working Group 10 developed and published the ISA/IEC 62443 Series. These documents describe a methodically engineered approach to addressing the cybersecurity of IACS, and the papers can be purchased from either organization.

Many other initiatives can be used as the base standard and best practice to rely on designing the best resilient cybersecurity solution for OT. However, you need also to follow the local regulations and required compliance to be considered.

For example, recently, Saudi Arabia, the Kingdom that is the primary source of oil export in the world and other petrochemical industries, has published guidelines by National Cybersecurity Authority to address Operational Technology Cybersecurity Controls (OTCC-1:2022), so if you are working on proposing a security solution for a food manufacturing or other means of OT within KSA, you have to follow these guidelines.

Additionally, there have been many initiatives from the US government following political conflicts and reported incidents of cyberattacks on the critical infrastructure within the US to invite commercial cybersecurity vendors to work together in fighting against these cyberattacks, which is quite interesting to see competitors join forces for a good cause, and that's the one thing unique in OT Cybersecurity that we're all working with the good side to defend the lives of people and social life order.

How to understand these standards and guidelines?

Usually, these documents are not easy to deal with; they give you the impression that lawyers have written them. This chapter will try to simplify the process to the maximum and explain the guidelines in more accessible language. Our leading example will be IEC 62443.

We shall begin by defining and simplifying the different parts and gradually put that in a context that will make sense at the end.

IEC 62443 Goal

"The primary goal of the IEC 62443 series is to provide a flexible framework that facilitates addressing current and future vulnerabilities in IACS and applying necessary mitigations in a systematic, defensible manner."

What are the elements of any standard or guidelines to which it's mapped?

People, Process, Technology (**PPT**). The PPT framework applies to standards and guidelines, where it helps identify how the three elements interact with each other.

They are the main three pillars of any organization; if the change impacts any of the three, it will cause an imbalance in the organization, which indicates that a change must also take place on the other two to achieve balance.

The **people** are the human resource in the organization; onboarding the right people who have the proper skill set and capabilities to perform the tasks is essential for having an efficient business, which also makes it necessary to assign the appropriate **roles** and **responsibilities** to the proper **function** within the organization.

People must adapt to processes and technologies; if they reject a process, it will become useless, or if they don't see the value of technology, then they are not going to use it, so people are addressed in the positioning of any new process or technology; it has to fit with their culture and work environment, they need to see values and how it improves achieving organization objectives or simplifying their work.

Processes are the workflows and steps taken to achieve a goal; they define how people can do their work. if an employee is requested to perform a task, an approved process must be followed.

People must understand processes, be trained on the process, understand their role, where it fits in the process, and is the objective to achieve as an outcome of the process.

Processes must also be measurable; clear metrics must be defined to decide on the status while in progress and when done.

Technology is the tool that helps people efficiently work through the process, and while we need to keep in mind that technology will only improve clearly defined problems, we don't acquire technologies because of an excellent sales pitch or luxury look; they have to serve a purpose of either improving the efficiency of solving a known problem.

The three pillars can't be separated; they have to work in harmony; people must understand the process and technology, what problems they solve, and what goals they achieve to have a balanced and efficient organization.

Role's definitions

Asset owner: The end-user, an individual or organization that owns and operates one or more operations.

System Integrator: The party who builds the operation and is responsible for the testing, implementation, integration, and maintenance of the different hardware and software from other OEM vendors.

Product Manufacturer: The OEM vendor who designs, builds, and supplies the different components of the system.

In project management, there's the concept of **RACI** to define responsibilities. The project is split into a list of deliverables and establishes the level of involvement of each party. They are as follows:

R: Responsible for doing the work and delivering the task.

A: Accountable, delegates the task, and performs a final review to confirm task completion.

C: Consulted, impacted by the task, and must provide input on the deliverable.

I: Informed, in the loop, and always informed of the progress.

By identifying the roles, you can easily use RACI or other techniq uesto define the different responsibilities of the various parties.

When following specific standards, it will be as a reference for the different roles but in different meanings, for example, in IEC 62443:

- **Asset owner**: They use it as a reference for cybersecurity objectives and assess the operation's current security posture.
- **System integrator**: Easier to understand standard cybersecurity requirements and have the team trained and ready to deliver and map the various system capabilities to the standard.
- **Product Manufacturer**: Have a baseline of accepted security requirements within the design of their products and can be used for the commercial competition of add-ons and differentiators.

The standards will also define general domains and sub-domains; for example, the IEC 62443 has the following sections:

1- General (62443-1)
2- Policies and Procedures (62443-2)
3- System (62443-3)
4- Component (62443-4)

Each of these sections has sub-parts, as follows:

General	IEC 62443-1 General Information	Concept and models	Master glossary of terms and abbreviations	System security complinace metrics	IACS security lifecycle and use cases	
		Part 1	Part 2	Part 3	Part 4	
Asset Owner	IEC 62443-2 Policies & Procedures	Security program requirements of IACS	Security protection rating	Patch management the IACS environment	Requirements for IACS service providers	Implementation guidance for IACS
		Part 1	Part 2	Part 3	Part 4	Part 5
System Integrator	IEC 62443-3 System	Security technologies for IACS	Security risk assessment and system design	System security requirements and security level		
		Part 1	Part 2	Part 3		
Product Manufacturer	IEC 62443-4 Component	Secure product development lifecycle requirements	Technical security requirements for components			
		Part 1	Part 2			

You can notice the mapping between different series and roles. The general information is to establish concepts and models, provide a glossary of all terms in use, and then specific parts for the three different roles.

IEC 62443 defines Security Levels (SL), which vary based on required knowledge and motivation; the higher the lever requires more measures to mitigate the risks.

The security level is measured using attacks (human error, simple, complex, sophisticated), and the motivation would indicate if it was a simple insider or intruder attack with little knowledge and resources or it's an advanced attack with extensive resources that hacktivists, cybercriminals, or state-sponsored attacks might lead.

SL	Means	Motivation	Resources	Knowledge	Protection against
0	-	-	-	-	No special requirements
1	-	-	-	-	Unintentional or accidental misuse
2	Simple	Low	Few	General	Intentional misuse
3	Complex	Medium	Moderate	OT specific	Intentional misuse
4	Sophisticated	High	Extensive	OT specific	Intentional misuse

The use of the security levels depends on the role. Product manufacturers will use the security level to identify the security functionality within the product, whereas an asset owner would determine what the security requirements are.

What we understand from the security levels is that it sets the level of expectation, for example:

- SL0 means the product has no security functionality, and the asset owner won't have any security requirements.
- SL4 means the attacker has very advanced capabilities and knowledge of the OT (for example, recent attacks on Ukraine using Industroyer 2). The attacker (adversary) could be state-sponsored and resourceful, so the asset owner would expect the product to have the functionality of protecting against this type of sophisticated attack. Hence, it will imply that the product manufacturer implements such built-in protection, and the system integrator will need to implement security controls to complete the security gap where product functionality is not fulfilling.

As a result, we observe the security levels as security requirements to mitigate specific weaknesses. These requirements are defined in IEC 62443 as follows:

- IEC 62443-1-1: Foundational Requirements (FR) defines **general** security requirements.

- IEC 62443-3-3: System Requirements (SR) defines the **whole system**'s security requirements for the system integrator.
- IEC 62443-4-2: Component Requirements (CR) defines specific product/**component** security requirements for the product manufacturer.

This is also an easy way to define responsibility: If the issue is related to the whole system -> it's the system integrator, and if it's associated with a specific component -> it's the product manufacturer.

What is the relation between SL and requirements?

When assessing an operation, you will need to determine the target Security Level (SL). Accordingly, you will identify the FR, SR, and CR requirements to address based on the target SL.

Whereas General Information – Concept and Model (IEC 62443-1-1) covers people, processes, and technology to achieve the required IACS Security Level, the System – "System security requirements and security level" (IEC 62443-3-3) defines the technical requirements for the IACS security level.

What are the FR Requirements?

For the Foundational Requirements (FR), IEC 62443 defines seven requirements:

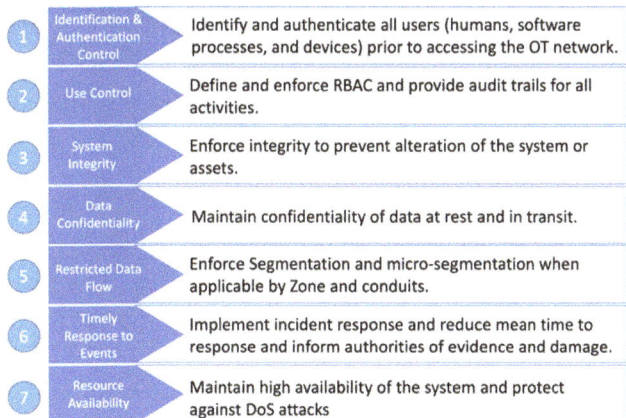

Each FR has a list of sub-requirements; for example, FR1 has parts such as wireless access management and multi-factor authentication.

The SR and CR will be mapped and aligned to the list of FR, so as per the standard of the IEC 62443 documentation, for each security level, there will be a matrix of FR that must be matched in the SR and CR.

Example of matrix FR1: Identification and Authentication Control:

Requirements	SL1	SL2	SL3	SL4
SR 1.1 – Human user identification and authentication	✓	✓	✓	✓
SR 1.1 – RE – 1 – Unique identification and authentication		✓	✓	✓
SR 1.1 – RE – 2 – Multifactor authentication for untrusted networks			✓	✓

- *RE: Requirement Enhancement*

This will help you and the asset owner, after defining the security level (SL) to identify precisely what requirements from the system integrator are to be implemented to the system and, similarly, what the product manufacturer's needs are to be available in the component.

How to use IEC 62443?

First, you need to purchase your copy; the softcopy might cost approximately 400$, but you need to check for yourself how much it costs when you intend to buy; it can be purchased as a single copy, which means it will be licensed under your name and can be a soft copy with the option to print it once or it can be a hard copy; you can also buy multi-user copies, where the price will be different; it's your reference, and it's worth the investment either for business or for educational purposes. Let me be clear: There's no referral program or any commission I will get if you buy it. I hope this has been established. ☺

The document will provide you with extensive details that can't be found in this book, online, or on any material except through purchasing a copy.

For the system integrator, they will use the document to define the security zones and the target security level (SL-T).

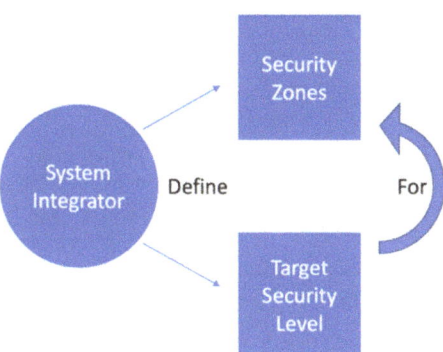

Product manufacturers will use the document to understand the components' requirements for a specific Security Capability Level (SL-C), where requirements might be natively available within the part or integrated with another system.

CR is derived from SR, which is derived from FR, where CR might have additional enhancement RE, which eventually defined the Security Level Capability (SL-C).

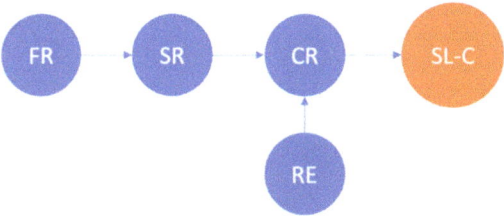

Throughout the document, you will find the mapping between FR requirements and SR/CR requirements; numbering will match in most cases, with some exceptions.

Defense in Depth

It is a concept in which multiple layers of security controls (defense) are placed throughout the system. It intends to provide redundancy if a security control fails or a vulnerability is exploited that can cover aspects of personnel, procedural, technical, and physical security for the duration of the system's life cycle.

This concept implies that an appropriate control will be implemented for each defined security layer.

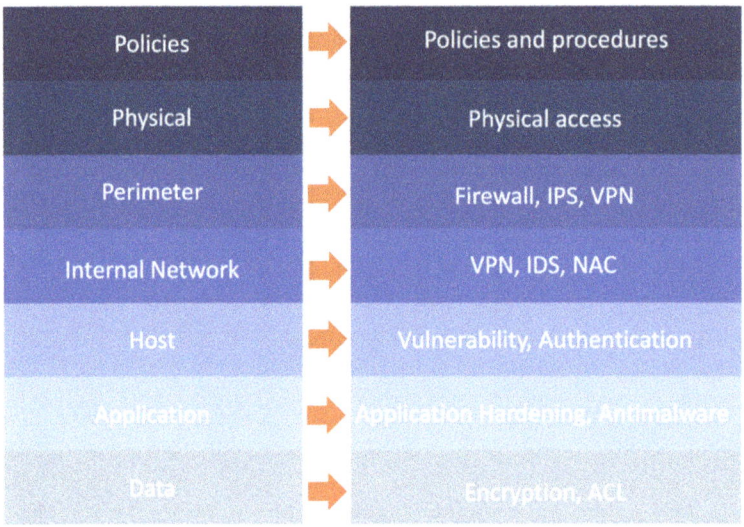

In IEC 62443, it defines four types of component requirements CR:

- Software Application
- Embedded Device
- Host Device
- Network Device

Zones and Conduits

These two terms are similar but different; unfortunately, I have heard some experts speaking about them as if they are the same, which is not the case.

These terms are related to networking infrastructure, and earlier, we have touched a bit on the Purdue Reference. Still, we need to distinguish here that Purdue Reference will identify where the assets belong, but it doesn't define what networking devices and L2/L3 devices are implemented.

Let's start from a flat network, where all the devices are on the L2 network and have no L3 boundaries, and you should advise whether to segment using L3 or implement proper controls.

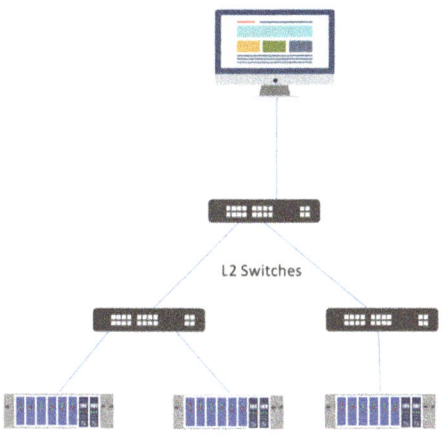

You may mistakenly hear that there are no actual controls for flat networks except through MAC address bypass. Still, different controls can be used to secure the communication within the flat network, such as L2 NGFW, which exists but won't refer to any commercial product within the book. The same is true with modern NAC, which can achieve Network Access Control use case without the need to have agent able devices and won't rely only on MAC address but instead on device classification based on several properties, where both controls can also enforce segmentation and micro-segmentation.

Zones

It consists of grouping cyber assets (physical or virtual) that share the exact cybersecurity requirements. They must be segmented either physically or logically on the network.

They can be defined as per plant; each process will form a different zone; or based on device type and function, for example, a PLCs zone.

In IEC 62443, where SI must initially define the security zone, they have to maintain a sound understanding of the plant and processes to determine the different security zones within the operation.

How are the zones segregated physically and logically?

Physically, by implementing an L3 device to route traffic between zones, each zone will be connected to the central L3 router/switch.

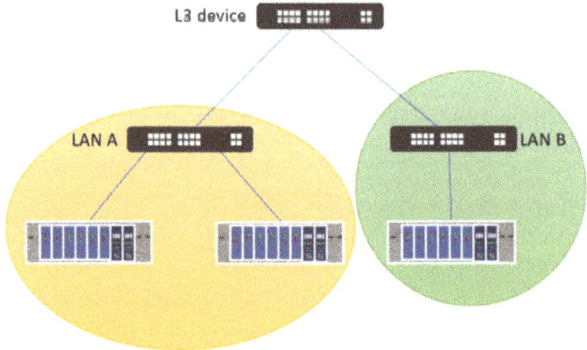

Logically, using VLANs within the same network device (switch), split the zones into virtual LANs.

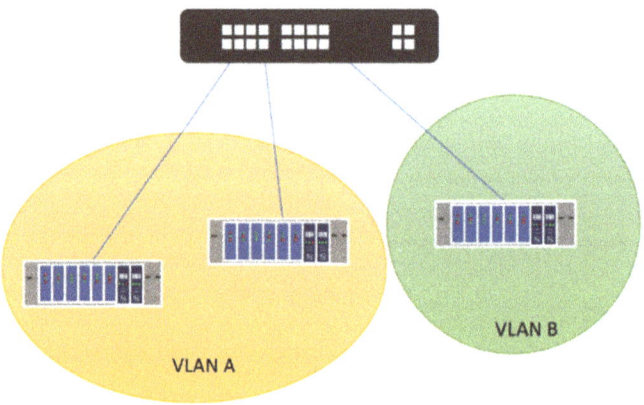

Essentially, zones are concerned with how to have the cyber assets with the exact security requirements communicating within the zone and define boundaries (VLAN, Router) for exiting the zone.

Conduits

Like zones, it's about grouping cyber assets that share the exact cybersecurity requirements, but are different from zone because it requires implementing security controls and properly segregating these groups into conduits.

Instead of L3 router or VLAN, we will have NGFW and DPI or modern NAC controls. These controls can enforce segmentation between conduits and block all unwanted communications while allowing only authorized communications.

This is an excellent approach to containing a compromise in part of the operation, by proper segmentation will prevent the infection/damage from reaching other parts of the operation.

But that also requires a deep understanding of the communication protocols, where your expertise and knowledge that you have obtained in previous chapters enforce appropriate traffic filtering and the creation of conduits.

Zones are a grouping from a networking perspective, while conduits are grouping from a security perspective, which means you may have multiple conduits within the same zone, but not the other way around.

Cybersecurity Maturity Model

The model is used to assess an organization for its cybersecurity posture level. This concept is obtained from the Cybersecurity Maturity Model Certification (**CMMC**). **In version 2.0, an assessment framework and assessor certification program are currently** designed to increase the trust in compliance measures to various standards published by the **NIST**.

We are concerned about the concept of the maturity model because it sets the foundation baseline for an organization's cybersecurity posture reality, and it can help design the strategy to achieve a targeted cybersecurity posture, or, as we can call it, the maturity model.

The model will provide guidelines of what to be assessed within the organization in terms of people, processes, and technology and, as a result, will decide which level the current posture is.

There are 5 levels:

1- Initial
2- Developing
3- Defined
4- Managed
5- Optimized

The assessment of the security posture of an organization's PPT will provide a result of which level the reality of the organization is at, following an example of a general description of the posture for the different levels:

It's similar to the IEC 62443 Security Level, which will identify the requirements for systems and components in the Maturity Model, which can help define the strategy of the required area of improvement to have a better security posture.

The most crucial point at the end of this chapter is that even if the operation is entirely compliant with IEC 62443 and the cybersecurity posture is at level 5.0, that doesn't mean it's cyberattacks' proof, as compliance doesn't necessarily equal security.

Compliance is from an accountability perspective. The security requirements are met, yet it can still fall victim to a cyberattack. However, compliant operations can mitigate the risk and handle the incidents better than non-compliant ones since they have controls and processes that defend and reduce the damage and can recover faster.

Standards would help as a baseline to make sure the overall security posture has been checked, but understanding the operation in detail, assessing the vulnerabilities and weak points, and addressing them with proper controls and processes also matters.

Chapter 13: Risk Assessment

Any new proposal, either a new strategy or implementing security controls, or any changes to improve the security posture, must be justified and based on findings.

A common mistake is to assume that every must-have operation list of security controls is implemented regardless of the type and details of the operation, so our proposals and acceptance from the Security team must be based on the reality of recognized needs.

Each security control is supposed to mitigate/reduce specific risk(s), and if the control is not aligned with the risks, then most probably, it will be missing the purpose.

We need to identify the risk and understand its impact, because if you implement an NGFW to provide segmentation within a security zone to create two conduits, that is not enough to say we have aligned a control to the risk. After all, we need to know what are the ACL rules and protocol DPI is activated on the NGFW to make sure they are aligned with the risk.

Risk is related to weakness or vulnerabilities in the system and linked to threats and impact. How do we distinguish between these terms, and how are they associated?

What is Vulnerability?
It's a flaw or weakness in:

- the system
- its security procedures
- internal controls
- design and implementation

which could be exploited to violate the system security policy.
<NIST SP 800-28 version 2>

How often your personal computer or mobile phone asks you to perform an update? That's precisely the mandatory security updates. The OS has identified some weaknesses that can be exploited and issued a fix, which could sometimes be highlighted as urgent/critical.

Software applications regularly publish updates to patch/fix newly detected vulnerabilities; it's only limited to software, but also hardware vendors might publish fixes/workarounds for some detected vulnerabilities.

Some vendors implement hidden vulnerabilities (backdoors) in their products for maintenance and the ability to access and recover in some cases, but might be forced by the government in other cases.

In the OT world, most assets are vulnerable and not patched; this is a fact you must realize because most of the components and communication protocols are insecure by design, which we have discussed in earlier chapters.

The obsolete product (software and hardware) don't get fixes and patches anymore after the end of support, which means hackers could discover a new vulnerability in the latest version of the product that has been made publicly available along with a fix/patch, what they could do is to test if the old/obsolete version of the product is also vulnerable, and if yes, it means they can easily exploit it because they know it's not getting support from the vendor anymore.

What is scary about vulnerabilities are those discovered by attackers and not known to the vendor (they are called zero-day), those that have been rejected by the vendor or didn't give enough attention, with a bounty hunter reporting them asking for a reward.

There's also the risk of a time gap between discovering a new vulnerability, the time it takes to be reported, and the time required to develop a patch that doesn't impact the product's performance or functionality until it's available for end-users to apply. The time it takes end users to download these patches and use them in the system. Those who are following the correct procedures will need to test it first in a lab, and then once confirmed, they will apply it to production; this is a very long process that might last for a couple of weeks or might be longer. To the extent of months!

In OT, applying patches is not that easy; vulnerabilities must be analyzed in terms of likelihood, impact, and justification of patching. They need to be tested first, and once it passes the test, there will be scheduled maintenance to apply the patch.

These delays and procedures are in place for a good reason, but they also expose the operation in the meanwhile to the risk of exploiting these vulnerabilities until they are patched.

These examples and scenarios are the reality; at this point, what do you think? You might be thinking this is the highest risk, and it must be the top management priority to patch any vulnerability as soon as it's discovered. But there's another reality: we think differently according to our role and function.

The CISO is a manager and not a security professional; CISO manages teams, resources, budget, meetings with the board, business objectives, and many other tasks. While hackers think differently, they see weaknesses they can exploit; they don't wait for budget approval or a maintenance window of time.

You are trying to propose a solution to secure an operation; you also think differently; you see what can be exploited and would suggest a way to mitigate it by looking best solution in the market and request it, not caring about the cost, the support SLA, the maintenance contract, or any other operational details.

A financial manager who has the budget will request a justification, and they might be right sometimes; why invest 1M$ to purchase a solution to protect a system that costs 500K$? This might be argued that the reputation and data loss could cost 10M$, so the finance team will consider it, or maybe there's a recovery solution that costs 100K$ that can bring everything to be normal instantly, even if an attack causes severe damage, as long as it can be restored in a timely fashion that fits business objectives then it should be enough. So it's normal to find CISOs accepting vulnerability risk in some cases because they know their business and they are the one accountable for it, not you!

What is a Threat?

It's a malicious act/attempt on the cyber system by exploiting a weakness/vulnerability to gain unauthorized access, causing damage, or stealing sensitive data.

The attacker who does the malicious act is called a **Threat Actor**, and the path the attack takes to bypass security controls and cause the damage is called **Threat Vector**.

So the threat is always linked to vulnerability; a threat is an attempt to exploit a vulnerability; it's not always a bug in the software; it can be a wrong firewall rule that is considered the weakness, and the threat actor will exploit this weakness to get past the firewall to access the internal resources.

In risk management, other types of threats are to be considered:

- Adversarial threats, any intentional attack by a threat actor, could be a negligent employee, a contractor, a hacktivist, or state-sponsored.
- Natural disasters, such as floods, earthquakes, and fire.
- System malfunction, systems that don't function properly could cause damage.
- Human error, unintentional misuse.

What is Risk?

The risk is the probability of exploiting a vulnerability. In other words, what is the likelihood that a vulnerability in an operation could get exploited and cause damage?

Let's assume a PLC is vulnerable to a list of known vulnerabilities. Still, this PLC is isolated from the network and has an embedded HMI and controlling simple I/O, so the likelihood of this PLC being exploited is zero because it's not connected to the network. As a result, there's no risk in handling this case.

In another example, we have discovered a vulnerability in a PLC communicating Modbus protocol. This vulnerability, if exploited, will give the attacker remote access, considering Modbus protocol is insecure, and an attacker can gain remote access anyway, so this risk is accepted. We think about other countermeasures to prevent an attacker from reaching the network rather than patching this vulnerability.

These two examples are what we do in risk assessment.

What is Risk Assessment?

Risk assessment is part of risk management, which we will discuss later, in risk assessment it's the process of identifying risks in an operation and then assessing the likelihood of the risks along with the potential impact.

Once we identify the probability of the risk and the impact, management can decide what to do with the risk, as it may be prevented, mitigated, reduced, accepted, ignored, or transferred.

Risk assessment has to be aligned with business objectives; pure cybersecurity point of view is not enough to determine the risk, as it has to be evaluated within the context of the business objectives.

Risk Calculation

How do we calculate risk and give it a value? Risk is likelihood or probability, so we can calculate risk and end up with a percentage of 0–100%; we also need to categorize these values as Low, Low-Med, Medium, Medium-High, and High.

To determine the risk, we rely on the likelihood and consequence factors.

		Impact				
		Insignificant	Minor	Moderate	Significant	Severe
Likelihood	Almost Certain	Low - Med	Medium	Med - High	High	High
	Likely	Low	Low - Med	Medium	Med - High	High
	Possible	Low	Low - Med	Medium	Med – High	Med – High
	Unlikely	Low	Low - Med	Low – Med	Medium	Med – High
	Very Unlikely	Low	Low	Low - Med	Medium	Medium

We can have a numeric representation of the risk by using the following formula:

Risk = Probability X Impact

There are many different ways of representing the risk; it depends on the risk framework used and the type of business/operation. Let's assume that a manager would like to know the risk in terms of $ value for a particular vulnerability, say it's email phishing. In this case, we won't use likelihood but instead a statistical probability.

To do so, we check how many times an email phishing attack occurred in the past and how many of these phishing attacks targeted similar businesses/operations; we found that in the last year, there have been 30 attacks, whereas the year before there were 10 attacks, so in one year, the number increased by 300%.

Furthermore, we found that the report was based on surveying 100 similar organizations. Next year, we expect between 60 to 90 email phishing attacks, and we can assume 75 email attacks as an expected average this year (60+90)/2 = 75.

75 out of 100 organizations is 0.75 or 75%

Now about the impact, we also need to identify the effect of the attack; reports usually show the loss in $, and let's say the attack resulted in a 1M$ loss as an impact.

Then we can conclude that the email phishing risk in $ value this year will be:

0.75 x 1M $ = 750K $.

Don't forget that in the end, these are all speculations; they only give insights to managers to decide how to react to the risk.

For the same example, the proposed email security and employee awareness program against email phishing costs 50K $ every year; the manager, in this case, would justify this investment because the numbers show significant savings.

This exercise that we have performed in the last example is how we define risk management, and it's always a good practice to have a cybersecurity strategy and acquisition of cybersecurity controls aligned with risk management.

It's an overwhelming task for an individual or small team to perform all the risk management independently. There are products and services dedicated to helping, which we will discuss in the cybersecurity controls chapter.

Chapter 14: OT Cybersecurity Controls

How to protect OT from cyberattacks? This is the question we're trying to answer in this book: What are the controls and countermeasures we need to implement to defend against cyber-attacks proactively!

But let's establish a foundation for distinguishing between controls to be implemented to provide real-time detection and prevention from the controls that are mainly classified under response.

The response usually comes with controls such as OT SoC, Incident Response IR team/service, Playbooks, Endpoint Detection and Response **EDR**, and Threat Hunting. These are not covered in this book, where we are focused on the implementation of detection and prevention controls that we will discuss in this chapter.

People

The human factor is the most critical. General users must be trained on cybersecurity awareness programs, and engineers in the field must obtain more than common knowledge to understand risks and threats in their operations.

Having team members dedicated to OT cybersecurity, as it's the case for IT, is essential, and I have witnessed the birth of the OT cybersecurity department for many customers, although the path took many turns. Eventually, it's happening.

One of the challenges I saw is with the culture: Some engineers didn't feel the need to know about cybersecurity, assuming they have an air-gapped network. We have discussed earlier that there's always a link in reality. Hence, the awareness programs help raise the common cybersecurity practice of average users and give the engineers a sense of responsibility for their operations from cyberattacks, which usually work.

Security awareness programs are essential, and learning more deeply about OT cybersecurity, as we cover in this book, is the ultimate goal for OT cybersecurity professionals.

The idea is simple; there must be someone in charge of the OT cybersecurity; if there's no department/team for the same, you need to recommend that or manage it by IT Network, Application, or Security. You need to suggest some education on OT Cybersecurity for those from IT backgrounds.

Then, you need to recommend regular awareness programs; these come in different formats; mostly, they will be in terms of Videos, which employees have to watch, and marked as completed only if they watch all the videos and answer assessment tests at the end. And it's repeated twice a year, for example, or on any suitable schedule.

There are other formats to run a campaign on the internal team and test if they fall victim to any known attacks, such as phishing, and then provide a program accordingly to follow to ensure they don't make the same mistake again.

An objective of accountability is also reached through this training because employees will be held accountable for any misconduct.

Obviously, this has to be discussed with C-level and HR departments as an essential control to be implemented.

Visibility and Classification of Assets

An asset owner must maintain an inventory of all assets, with classification details, the role, function, network information, physical location, logical grouping, vendor, model, serial number, and criticality of the asset. In an earlier chapter, we have taken some examples of how to build a sheet with a list of support, which is mandatory to be established and updated regularly.

The information that is built must be tracked for changes; every change MUST be justified or investigated. Suppose you had a device listed as a PLC, and through an update, you found this device communicating as a Raspberry PI; this is a highly suspicious change and requires immediate investigation.

Is it possible that an asset owner doesn't know about the assets within the plant? Most probably, yes; when you have large plants and operations of thousands of distributed assets, it's tough to maintain complete visibility of all the assets, and classification of the assets is even more complicated.

The inventory of assets will form the **assets baseline**, which can be used to compare any new changes and alert on findings, but I hear you; it's a challenging manual task!

Although site surveys can always build information on the assets, it's not enough, and visibility solutions must come into the picture.

The most suitable solution to be used is the OT Intrusion Detection System (OT IDS). However, the name indicates it's dedicated to intrusion detection. The main values are visibility and classification.

Intrusion Detection System – OT IDS

In some cases, it's called OT network monitoring and has two approaches:

- **Passive**: Where it relies on network-mirrored traffic to the IDS to perform the analysis and extract information from network flows.
- **Active**: Not recommended in most operations; if required, it has to be controlled for the purpose of scanning certain IT assets within the OT and not pure OT assets.

Most solutions use the architecture of sensor – management, which means they deploy sensors at each plant or next to network switches in the plant to receive all network mirrored traffic. They might need several sensors to cover all plant switches. And they are connected through the OOB network to the management console, where they provide feeds and analytics.

Many challenges you might face when implementing an OT sensor, whether it has built-in management or function as a sensor only and is managed separately by the management console. These challenges are:

- Network switches don't support port mirroring (SPAN).
    - In this case, you have to recommend one of two solutions:
        - Use TAP devices.
        - Include new network switches within the scope of IDS implementation.
- There is no space in the cabinet to deploy an extra device (sensor)
    - Check the possibility of providing remote SPAN and place the sensor in the most suitable location that has space and can receive the mirrored traffic.

- o If not possible, you need to investigate other possibilities of creating space, adding a TAP device to provide remote mirrored traffic, or other alternatives.
- You can't mirror from all the switches; you have to select some switches for mirroring.
  - o Based on a sound understanding of the operation communication, you need to pick the suitable switches with the most valuable information and risk missing some assets.
- The sensor needs to be deployed in a harsh environment.
  - o Check the vendor hardware guide to determine which model supports the target environment.
  - o If a vendor doesn't have suitable hardware but selected because it fits other requirements, you can use remote SPAN or TAP and place the sensor in an appropriate environment.
    - Or if the sensor can be deployed as a virtual appliance, then propose HW that fits the needs and install proper hypervisor supported by the vendor, and make sure to have the physical NIC mapping to the Virtual appliance directly and not through virtual switches of the hypervisor, for example in VMware it's called PCI Pass-through.
- The operation communication bandwidth is higher than the sensor capabilities.
  - o Select the suitable sensor model; if not enough, propose multiple sensors.
    - This is very rare to be seen. Usually, OT communication throughput is low.
- The sensor has six interfaces, one for management and five for sniffing the mirrored traffic, but there are nine network switches to mirror from.
  - o Make sure to select a suitable model.
  - o Check if the vendor supports adding extra NIC cards.
  - o Check if virtual is possible, then build the proper hardware with Hypervisor.

What outcomes to expect from an OT IDS?

- OT Network visualization.
    - Different solutions in the market provide various ways of graphically presenting the OT network communication while showing the best values on the visual map.
    - They can show the assets and links between the assets; it's a nice view of overall assets and communications.
    - Some views provide mapping to the Purdue model to logically assign assets at the proper level.
    - Some offers drill down to the details of each asset, and then you can obtain all possible information about the asset.
    - Some may not be useful in large operations with thousands of assets, depending on the visuals. Still, they usually offer grouping and clustering to show the similar assets together and make a helpful view.
- Asset Baseline
    - The IDS will build an inventory of all assets, which could be:
        - Assets with IP address.
        - Assets with Mac address only.
        - Nested assets are connected behind a device (PLC, Router, etc.) and may be connected over serial or as a module and not directly communicate with the network.
    - Asset classification, where details about each asset are collected and viewed per asset.
    - Track changes on the assets; after determining that the **assets baseline** is accurate, the baseline building will be paused, and consider any new change as something suspicious, which can then send an alert to investigate or justify.
- OT Protocol Baseline
    - When deploying the IDS, it will start in learning mode, where it will record every communication on the **traffic baseline. Depending** on the solution's capability, it can support DPI for OT protocols and extract details about read/write commands on the network and consider it as part of the baseline or what defines a normal operation.

- o The learning process is risky because we assume that everything occurs on the OT network during the learning process as a normal process, which could last for a couple of weeks until we're sure we have recorded every possible communication on the OT network.
- o A process tuning this baseline later must be reviewed and ensure it reflects the normal process and no dangerous operations are included.
- o Once the baseline is built, the IDS will be configured to alert on any deviation from the baseline, so any new communication that was not recorded in the baseline will cause the IDS to send an alert.
- o You need to make sure the IDS solution provides DPI support for the OT network protocols, and if not, you need to request the support; as usual, they offer delivery of new protocol support within weeks.
- Intrusion Detection
  - o Each solution has a set of techniques to detect threats; the common is a deviation from the baseline. Some implement anomaly detection and maintain a database of OT/IT attack signatures to alert in case of detection.
  - o Some solutions offer Threat Intel Feeds, and it helps scan the network for known attacks and scan the logs backward and detect if any malicious traffic took place on the network in the past.
  - o You have to examine the IDS intrusion detection capabilities. You request some testing to be performed on the test environment or production if it's passive and can easily be implemented, or even using packer capture, as usually, IDS solutions offer the possibility to perform offline Proof of Concept (PoC) by replaying packet captured files.
- Vulnerability Scan
  - o OT IDS also can act as Passive Network Vulnerability Scanner, which means they can detect vulnerabilities on assets based on network flows.
  - o They should provide a kind of scoring per vulnerability, ex: The Common Vulnerability Scoring System CVSS.

- o Additionally, providing a description per vulnerability with resource links will help you understand the vulnerability and the risk you're dealing with.
- o Some solutions offer matching confidence because passive scanners are not as accurate as active scanners, so when the solution is sure about the vulnerability, it's different from when it's indicating it's probably vulnerable.
- o The solution must provide information about the impact; this is important to decide how to respond; similar to previous examples, not every vulnerability must be patched; it all depends on the impact.
- o A solution as either workaround or an official fix as announced by the vendor is required to provide engineers guidelines on how to fix it if required.

The OT IDS establishes the foundation of assets and communication visibility; its core value is about not being in line for the traffic; it should not interrupt communication. At the same time, it can see the traffic from all parts of the operation and is not limited to a segment, which gives the resulting visibility providing a complete picture.

Next-Generation Firewall NGFW

Firewalls works on L3 and L4; they can set access control lists (allow/deny) based on source and destination IP and port.

Traditionally, firewalls used to be stateless, which means they needed to have rules to allow the request in one direction and then another rule to handle the reply.

To allow access for an IP to a Web server, it was like this:

Allow: Source IP: X.X.X.X – Destination IP: Y.Y.Y.Y – Source Port: XXXX – Destination port: 80

This rule would allow the request from X.X.X.X to Y.Y.Y.Y

Where the reply will be in the opposite direction is, it needs to handle the response in the following rule:

Allow: Source IP: Y.Y.Y.Y – Destination IP: X.X.X.X – Source Port: 80 – Destination Port: XXXX

You can realize how risky it's because the source port will be a random range, and that means the Firewall needs to allow a range of ports to access the internal resources.

Then there is the introduction of a stateful firewall, in which, in essence, you will only need to create a rule for the request, and the firewall will handle the reply.

But the stateful firewall functionality is made available in almost every system, built-in in all OS, so if the purpose is only to enforce ACL in this way, any basic router can have the capability to enforce the ACL rules.

The limitation of the firewall to only filter based on L3/L4 has been recognized in the past because of the possibility to tunnel different applications through any port you may desire, besides many basic techniques that can bypass the firewalls ACL, for example, IP address spoofing.

What was required was to understand the upper layers, not only L3/L4. It was needed to understand the application level, which was possible through implementing network IDS, which can analyze the payload of the packets and inform the firewall to block the connection if anything is suspicious.

An improvement to IDS resulted in introducing a new solution called Intrusion Prevention System **IPS**, which can investigate the packets and terminate the connection and not only alert.

It didn't stop here, as later on, some security vendors integrated IPS and Firewall into one solution and called it Unified Threat Management **UTM. With** time, they added more capabilities such as network antivirus, antispam, and URL filtering, and the competition started around adding more and more features to the UTM.

After UTM, the Next-Generation Firewall **NGFW** was introduced to the market, which is very similar to UTM since it has multiple network security engines, but with a focus on applications. Because of the shift from web sites to web applications, they needed security solutions can identify the applications communications' regardless of the port number in use.

The NGFW, through its embedded IPS, started to include DPI support for some industrial protocols. However, this support is limited to a few protocols 10–20 and can't be compared to OT IDS, which usually covers between 100–200 protocols. As a result of this gap, it's common to see commercial integration between NGFW and OT IDS to provide complete detection and prevention for OT protocol attacks.

The NGFW will be used to create security zones, conduits, DMZ, and separate different levels; for example, a firewall will be implemented between layers 3 and 4 to create a layer 3.5 DMZ hosting all security services and isolating the lower levels from business and the internet.

It can also be used as a segmentation point to isolate different conduits from each other and only allow specific administration traffic.

When considering NGFW, you need to check for the following:

- Support of OT Protocols: Although we rely on OT IDS, it's better if the NGFW supports the OT protocols for better security and to achieve defense in depth.
- Hardware compliance with environmental needs or the choice to switch to a virtual NGFW running on supported HW and hypervisor.
- The number and types of interfaces.
- The required throughput.
- The possibility to manage multi segments, which enables the NGFW to be a central point between multiple segments and has the capabilities to manage them, sometimes it can be supported out of the box, or it has to use the concept of virtual domains, where the physical NGFW can run multiple virtual NGFW each will handle a separate segment. I am aware of some products that can run in L2 Bridge while performing complete L3 functionality; it will be helpful to have the NGFW connecting several networks as an L2 bridge while enforcing all means of control.
- SD-WAN support if required, some industrial switches support SD-WAN today and can integrate with the fabric of SD-WAN; if it's a new plant and has the SD-WAN implemented, you need to make sure the NGFW can fit.
- NGFW is expected to handle integration with the network infrastructure in routing protocols, VLANs, 801.q trunk, link aggregation, and others that are also required to fit it.

Endpoint Protection EPP

Traditionally Antivirus! Which changed through different stages as Antimalware, Host-Based IPS, and Application Whitelisting, until it's now under one category of Endpoint Protection EPP.

Don't confuse EPP with EDR (Endpoint Detection & Response)!

EPP is an Agent-Based, and as it's the case of every Agent-Based solution within OT, there's a common requirement: It's not up to the asset owner or system integrator to decide which agent can be deployed; it must be tested and certified by the product manufacturer, and that's a mandatory requirement.

What functionalities to expect from the EPP?

- Scan endpoints for known viruses and malwares.
- Detect malicious files.
- Support offline updates, where the management of the EPP should not be connected to the internet, it must provide a possibility to download the updates and upload them to the EPP agents offline; this includes documented procedure on how to:
    o Download the update files.
    o Verify the hash of the downloaded files.
    o Secure transfer of the files from the corporate network to the OT network.
    o Upload the update files to console.
    o Push the updates to all agents.
    o Verification of an update on the agent side.
- Provide a Memory Intrusion Preventions System **MIPS** that can detect all types of memory attacks, like injection and buffer overflow, to prevent them.
- Secure system drivers, as infecting system drivers, will corrupt many functionalities.
- Security for the OS kernel, a dedicated inspection for internal system calls between OS and user land, is not available in many solutions, but it plays an essential role in preventing 0-day malware.
- Host-Based IPS (HIPS), similar to Network IPS, a HIPS can detect and prevent attacks from the network to an asset.
- Application whitelisting, a widely used technique in OT in general, limits any workstation to running only a list of approved applications and prevents everything else. Must be tested by OEM vendor and supported because they need to ensure it doesn't prevent applications required that were not part of the test scope. Additionally, the whitelisting cab is based on:

- o The application executable path is not secure because it depends on the file name and path, that can be altered.
- o A certificate of a signed application it's good to control, but unfortunately, you might not find that all used applications are signed.
- o Hash of the executable is the best accurate method because a hash of an executable will not change; however, maintaining it could be difficult since every update to the application will change the hash, so it must also be taken into consideration to update the hashes in the whitelisting.
- o Combining the path, filename, and certificate would also be good.
- File extensions linked to Applications whitelisting are not widely used. Still, it's a firm control against ransomware, this feature will link files of certain extensions to be only accessed by specific applications, so when you linked files of ".Docx" to be accessed by MS Word only, even in case of successful exploitation of the device took place when the malware tries to access the files to encrypt them it will be prevented. As a result, only OS and applications are impacted but not the files, so this could avoid becoming a ransom victim.
- Avoid cloud analysis; some solutions have extensive detection capabilities, but they rely on cloud analysis, which is not applicable for most OT networks to have a bi-directional connection with the Internet cloud, so if the solution becomes limited with no cloud access, it is not a good fit in this case.
- OS Hardening, although this in many articles sounds like manual work to be performed by a system integrator, it's an option that some EPP solutions can offer. And as indicated earlier, changes to Endpoints must be tested and approved by OEM vendors; if you remove unused libraries from the system and apply some controls on the OS/Applications, it will harden the system while following the proper guidelines (Linux/Windows/Mac). Still, you're risking the support void from the OEM vendor, so it's better either way, by an EPP or through a manual application, to obtain the OEM vendor approval.
- External Media: By realizing the importance of using external media, for example, the use of USB mass storage to transfer updates and other files

to and from the OT network, there are essential guidelines to be followed:

- EPP must be capable of blocking all external devices connected through available ports (such as USB and Bluetooth) and only authorizing whitelisted devices.
- The authorizing of devices requires the EPP to identify the device type, vendor, model, serial number, and any other properties to make whitelisting based on several properties for better security. In other words, you can allow specific USB mass storage, where the type has to be mass storage and it must match the defined vendor/model and the device's serial number.
- EPP must have the capability to mount external devices for Read/Write only and not for execution.
- When using external storage, it must be scanned by an isolated machine, preferably having a different EPP than the one in the OT network and another OS.
- EPP must be able to encrypt/decrypt all data of USB storage, and only the encrypted data that can be accessed from the USB storage to prevent access to any additional data added outside the process.
- No mount for any media type other than the required external storage; if the type of USB device is USB Ethernet or 5G modem, it has to be blocked.
- EPP must have the capability to issue an authorization code to allow using of USB storage, which makes every single-use unique, monitored, and justified.
- For Windows OS, EPP should also be able to secure the Registry keys and system config files and, when needed and as part of the OS hardening to modify some keys to prevent unwanted behavior, such as the capability of EPP to prevent the machine from booting in safe mode, because usually, safe mode disables most of the protection controls on the endpoint.

A general approach is to request from EPP vendors and also for other controls to explain how they achieve the required capabilities; in some cases, when there's an RFP for a project to acquire a solution such as EPP, the

Resellers will try to answer to all requirements in a commercial way, trying to avoid saying we don't support or we don't comply. At the same time, they might be able to achieve the same objective with different techniques. And by requesting technical clarifications about how they can achieve the specific feature of capability, it's easier for the committee who are validating the proposal to assess whether they can technically fit the requirement or not, regardless of the marketing terms and buzz words.

Identity and Access Management IAM

IAM is about providing the required access to the designated users to specific system resources within the accurate time and duration.

The first element of the IAM is the identity store, such as MS Active Directory and LDAP, where all users' accounts and details are stored; the following are essential aspects:

- There must be a central user identity store for the OT network in the design, and it's different from the IT user directory.
- Authentication protocols for the selected user directory must be supported in all OT networks where users are required to authenticate (Kerberos, LDAP, Certificate).
- The directory must support TLS encryption, as no authentication with plain text should be allowed in the OT network.
- Users are defined and assigned to appropriate groups, where permission will be assigned to groups and not per user.
- The directory should support self-service, where users can change their account details and reset their password if required, while the allowed change should only be limited to non-functional properties; for example, users can reset their passwords can change their display name, phone number, but they can't assign themselves to groups or create different email addresses.
- Password policy should maintain complexity, where users can only select hard-to-guess passwords.
- Password policy should force the users to reset their passwords at least once every three months.
- The security administrator will be assigning roles and permissions per group and also can assign users to groups.

- The directory must maintain audit trails for all users' activities, such as:
    - Login
    - Logout
    - Password change
    - Profile update
- The directory must maintain audit trails for all security administrators' activities, such as:
    - Account creation/modification/deletion
    - Change in the password policy
    - Block/unblock the account
    - Roles/Groups creations/modification/deletion
    - Assigning permission to account
- The directory must maintain continuous backup
- The directory must have backup authentication servers to maintain 99.99% availability

The first process of IAM is where users identify themselves by presenting an ID. Usually, a "username," following that, users must prove their identity through an authentication process.

Authentication is where users present one of the following:

- Something they have such as digital card or certificate, SMS code, application PIN
- Something then know, such as a password, or the answer to security questions
- Something they are, such as fingerprint

Authentication could rely on a single factor of authentication, or multi-factor, such as password and fingerprint, or card and SMS code.

Once the user presents proof of their identity, it will be compared with the Identity Store to confirm the validity of the account and the provided information, and the user will be authenticated.

The authorization process takes place after the user is considered authenticated; the user will request the service, such as interactive login to a workstation, access to a shared folder on the network, or connecting remotely to a server; based on the group/role of the user is assigned, they will be granted/denied based on the given permission.

The Identity store must also maintain temporary accounts with a start and expiry date/time, used for contractors who have scheduled maintenance access to the system. This means they can only start working when the account activation begins and lose access when the account is activated and expires.

A clear process of provisioning new accounts, enabling authentication, providing authorization, assigning permission, and self-service until de-provisioning of the account must be clearly defined because it's important when the function of an employee changes to be released from old permissions and assigned new ones, while an employee leaving the company or underperformance monitoring, or due to suspicious activities then it's must be possible to know when to disable the account and remove all permissions, which defines the IAM lifecycle, as in the following illustration:

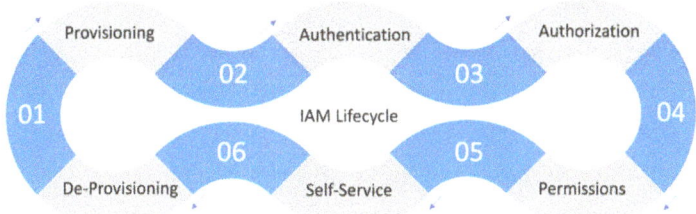

Single Sign on SSO

SSO is a nice feature that reduces the number of authentications required by the user since the user was able to log in to a desktop interactively. If the user would like to access any service on the network, SSO will play an important role in considering the user authenticated without the need for the user to log in again. However, it's not a top priority in OT networks, as it's the case for IT users.

Still, there are other types of authentications that users do not perform. Instead, they are performed by applications and services; to function, they require several accesses to the system, and they will be challenged by authentication; in this case, the SSO is required for applications and services to be able to function normally, but the segregation of roles is essential between what is assigned to services/applications from what is assigned to users; for example, services don't require interactive login!

Integration, although this is a general requirement for all controls, is critical that the identity store and authentication protocols in use are supported by other security controls and systems' components. Needless to discuss in detail, there's no use of an identity store that is not supported on the systems intended to be used.

How about authentication and authorization of devices, not users?

This is an important question when you consider devices that don't have users behind, such as printers, cameras, PLCs, and access doors, or, in other words, IoT and IIoT; how we can control access of these devices within the OT network?

The **N**etwork **A**ccess **C**ontrol (**NAC**) is the usual solution, with some considerations. NAC is known as the solution for device authentications and security posture assessment.

However, NAC solutions are not widely adapted in OT networks; for some reason, the focus of asset owners and systems integrators is on Network Management systems **NMS** and not on NAC. While there are many overlaps between the two solutions, and many experts see them as the same solution, let's discuss the solutions in a bit of more details to justify our recommendation.

NMS will provide monitoring for infrastructure devices; it will show if the switch is up or down, the status of ports, connected devices, throughput, and other monitoring information.

NAC can similarly provide the monitoring information of the infrastructure devices, but it extends that to devices monitoring, assessment, and control.

I have repeatedly warned about using any means of active techniques or scanning in the OT network because we're worried about the OT assets. However, network infrastructure devices are more tolerable for these kinds of scanning and remote management, so if NAC is in a place where it supports the monitoring and management of all infrastructure devices, that brings more value than NMS because we can use the NAC functionality to control access of IoT/IIoT devices within OT.

NAC traditionally was known to work in one of two ways:

- 802.1x Authentication
- **MAC A**ddress **B**ypass **MAB**

Whereas in 802.1x, it requires a supplication to be deployed on the device, or it has to be equipped with a built-in supplicant that can authenticate to the network switch or wireless AP it's connecting to.

In case the device doesn't support 802.1x or support deploying an agent, which is commonly the case of IoT/IIoT devices, then alternatively, it can be whitelisted based on its MAC address; it's weak but better than no control; it can't handle MAC address spoofing, where an attacker can simply spoof the same MAC address and connect on the network port of another device to gain access.

At this point, we still didn't achieve significant control on the network, so what is the suggested recommendation?

The Next Generation NAC is used to provide better access control on the network because it doesn't rely on 802.1x and MAB only; instead, it depends on authenticating several properties of the devices. To make it a bit clearer, here's an example:

NG-NAC can scan the network devices for properties, such as:

- MAC address
- IP Address
- OS
- Open ports
- Power consumption from a switch PoE
- Classification based on several fingerprinting

While it's not possible to deploy an agent, the NG-NAC will build the identity of the device based on a set of properties. If an attacker spoofs the same MAC address of a printer and connects a RaspberryPI, the NG-NAC will prevent access because it will see a match in one property (MAC address) while a difference in all other properties.

NG-NAC doesn't require an agent and works with all types of connected assets; it can perform in-depth posture assessment of IT devices such as laptops and also assessment of IoT/IIoT devices such as open ports and weak credentials.

Some cautions must be considered when deploying NG-NAC:

- Post-Connect assessment only, the NAC, in general, will quarantine the machine until it's authenticated. In OT, we can tolerate that, so the

device must gain normal access to the network and immediately be assessed if found suspicious enough, it will be quarantined.
- NG-NAC has comprehensive techniques of discovery, classification, assessment, and control; some tools have 30+ different techniques (WMI, NMAP Scan, SNMP, SSH); we have to distinguish which methods are allowed within the OT network; the rules will be as follows:
    o Passive techniques are allowed for all types of devices.
    o Active techniques are allowed to specific IT assets.
    o No active techniques are allowed on OT assets.
    o Active techniques are allowed to infrastructure in read-only mode.
    o Control actions must be completely disabled and used based on a manual trigger from the incident response team.
    o In general, the NG-NAC deployment plan must be identified, tested, and approved.
- Support, it's essential to confirm that the selected NG-NAC supports all infrastructure devices.

Secure Remote Access

Secure remote access is not based on a single solution; it's a combination of multiple solutions integrated to provide secure remote access to OT cyber assets.

It might consist of IAM, VPN Concentrator, NGFW, and other solutions such as session recording. It should apply the least privilege and complete auditing during the remote access session.

The IAM will be responsible for the authentication and authorization. In contrast, VPN Concentrator will be responsible for establishing a secure tunnel between the remote user and the OT network, while NGFW will be enforcing app-based ACL to allow very restricted access to the user to enable them to do the urgent work that had to take place remotely.

The IAM will manage the account lifecycle, while the VPN concentrator and NGFW will terminate the access once the account has been deprovisioned by the IAM.

A session recording SW will be used to record the complete access, and all previously involved controls must maintain complete audit trails during the remote access.

NG-NAC can play a role in assessing the security posture of the remote device prior to providing it access to the network.

The access should not be provided to a complete VLAN or subnet; it has to be provided to a dedicated Jump server; this server will be assessed for security posture and provide application-level access for the remote user and not network access.

All means of possible tunneling must be disabled; the Jump Server should not allow connecting remotely to another server through Remote Desktop. Another example provides access over SSH because it can be used to tunnel to other devices.

In simple words, based on an urgency that we learned about during COVID-19, we may provide remote access, but it has to be recorded, restricted, and controlled.

Alternatively, a better way is to use secure online meeting software, share the screen with a remote user, and make the work locally, not through remote access.

Zero-Trust Mindset

When designing any security policy and placing security controls for OT networks, you should maintain a zero-trust mindset, which means no trust for any user or device and assume there's an attacker in the network already. All your security policy should be about allowing the work to continue with no interruption and, at the same time, restricting the access for the hidden attacker in the network.

This is a complicated approach, but it's the right one. To help you with this approach, there are zero-trust solutions in the market; it's still a new concept, so not yet mature, but before implementing zero-trust solutions, adapt to the mindset.

Why am I claiming it's not mature yet while typing these words? Suppose you're looking for a zero-trust infrastructure solution. Then you find many claims in the market of solutions that can deliver zero trust, but from the nature of these solutions, they are not capable of providing a zero-trust! How is that?

In Zero Trust architecture, there are two different components:

- Decision point
- Enforcement points

The decision point should maintain complete visibility throughout all OT networks; it should maintain a policy for the overall network and not for a specific segment; this can be achieved by integrating with OT IDS, NGFW, EPP, NG-NAC, IAM, infrastructure, and all other components of the OT network. As a result, it has complete visibility of what is going on in the network and can assess the devices based on several factors and feeds coming from different solutions, which will make it possible to build contextual policies for the decision point based on several criteria and not based on a single solution, as a result when the contextual policy detects suspicious behavior on one of the devices it can alert, or if the matching confidence with a severe case such as Ransomware attack is 100% confirmed, it could decide to isolate the infected device and restrict its access.

When the decision is taken from the decision point, it will be enforced through the enforcement points; in this case, solutions such as NGFW can play the role of restricting the access of the infected device, the NG-NAC can quarantine the infected device, or the EPP can also quarantine it locally.

So when an NGFW or EPP claim to provide a zero-trust architecture, it's not possible in reality because they don't have complete visibility of the whole network, and that's why, until writing these words, it's not mature. However, several initiatives are progressing well. This might be the new control that can be adapted in OT networks with the rising complexity of cyberattacks; it's becoming necessary to see soon.

Kill Switch

In the last example of zero-trust, we wanted to have the ability to isolate infected machines, and we mentioned that for NG-NAC, the actions must be activated manually by the IR team. By these two points, suppose an attack took place on an OT network, and it had happened before the attackers took complete control over all PLCs/IEDs. The HMI was kind of read-only; operators could not control a thing. Instead, they were just watching what the attackers were doing.

What would be appropriate is to have the kill switch policies available and can be triggered manually by authorized users, a policy that can isolate complete supervisory devices from control devices, regardless of how they are physically connected; these types of policies are possible through the proper classification and control over the infrastructure, and the best candidate would be the NG-NAC.

A set of policies that are ready to use for different possible scenarios that we have learned about from previous attacks would save us and make the response and recovery much easier when we're in panic mode. Attackers are moving things around the OT network.

At the end of this chapter, I want to say that it's not a conclusive list of all possible cybersecurity controls within OT networks. Still, these are common ones, and the possibility to have others is there, but the point here is to see how we don't follow the marketing messages of cybersecurity vendors or follow the buzzwords. Instead, we put the control in the context of OT and define what is required and accepted and what is not suitable and rejected.

In principle, all controls must be tested and approved for OT; they must offer integration capabilities, such as API or common protocols. They also need to provide audit logs and trails and a real-time view of the status of devices and communications, the performance, and compliance; these logs must be collected centrally in a SIEM and must also be archived.

The SIEM is an essential solution; for the scope of this book, we only care about maintaining all the logs, providing a real-time view, archiving all the logs, and having easy-to-read dashboards, but further use cases of the SIEM and its capabilities fall under the IR.

References

https://instrumentationtools.com

https://www.NIST.gov

https://Cisa.org

https://Gartner.com

https://gca.isa.org/hubfs/ISAGCA Quick Start Guide FINAL.pdf

https://Mqtt.org

https://wiki.wireshark.org/IsoProtocolFamily

https://datatracker.ietf.org/doc/html/rfc905

https://BACnet.org

https://www.cimetrics.com/blogs/news/bacnet-device-objects-and-properties

www.ingramcontent.com/pod-product-compliance
Lightning Source LLC
Chambersburg PA
CBHW040520220526
45473CB00013B/2928